LANDSCAPE DETECTIVE

LANDSCAPE DETECTIVE

Discovering a Countryside

Richard Muir

For Catherine, Steve and Heather, as well as the other members of the Escape Committee yet to go over the wire.

Published by: Windgather Press, 31 Shrigley Road, Bollington, Macclesfield, Cheshire SK10 5RD, UK

Distributed by: Central Books, 99 Wallis Road, London E9 5LN

British Library Cataloguing-in-Publication Data
A catalogue record for this book is available from the British Library

Landscape Detective: Discovering a Countryside
by Richard Muir

ISBN 0 9538630 2 6

First published 2001

Typeset and originated by Carnegie Publishing Ltd,
Chatsworth Road, Lancaster
Printed and bound by Alden Press, Oxford

Contents

List of Illustrations

Figures

Colour plates

The colour plates are reproduced between pages 86 and 87

Introduction

Two main themes are explored in this book. Firstly, by discussing the techniques and reasoning involved in investigating real problems of landscape history and archaeology, I show the ways in which the landscape detective goes about his or her work. Secondly, a 'total' landscape approach is applied to an unusually interesting township, Ripley, in the Yorkshire Dales, providing a case study from which the value of the approach may be gauged.

Landscape detection

I think that I can claim credit for coining the term 'landscape detective' (Muir, 1981, 16). Some twenty years have passed since then and the notion of detection seems more apt than ever. Those of us who engage in landscape research find ourselves dealing with evidence of many types and origins. Some of it we can handle quite easily – like the clues contained in old maps or in (some forms of) earthworks. Some lies at the margins of our competence – this might include the detailed analysis of air photographs or the dating of fragments of pottery. Some may require expert advice, in this case this involved the dating of local medieval pottery. In all this activity, the actions of the landscape historian/archaeologist have a great deal in common with those of a police detective. Similarly, in many cases the evidence will be too thin to constitute watertight proof. We then find ourselves juggling with the probabilities and seeking best-fit solutions.

The question of what constitutes proof is an interesting one both for criminal detectives and landscape detectives. There may be a tendency to assume that an 'expert' with degrees as well as papers and books in print will be infallible. This one certainly is not. Indeed, when one looks at interpretations provided by 'great names' one may be amazed at how thin the evidence underpinning their conclusions can sometimes be. I try to be most critical of my own judgements and always to remember that the more that one becomes wedded to prematurely formed interpretations to the exclusion of other possibilities then the more one is likely to be spectacularly wrong. Interpretations can tend to change quite dramatically as the depth of analysis increases. This is rather worrying, for in any study there will be core issues for in-depth analysis and peripheral matters and context questions that will receive a much skimpier treatment.

FIGURE 1.
An oblique aerial view across the Ripley deer park, with the village and garden earthworks in the foreground. The very faint ridge and furrow running down to the beck to the left of the garden is clearly shown. Vegetation has rapidly encroached upon the further lake during the second half of the twentieth century. At the lower right margin pronounced broad ridge and furrow is seen running under the village.
© ENGLISH HERITAGE

Experience also shows it to be the case that if one becomes fixated on some phenomenon or idea then one will surely discover it in the field (whether it is there or not). Long ago, in satirising the mumbo-jumbo world of leylines (the Young Farmers had yet to discover crop circles) I decided to look for a mythical animal associated with King Arthur traced out in the lanes and boundaries of the Glastonbury countryside. *Arthur* was sometimes translated as 'bear' and literally within minutes a most convincing bear had emerged – of course it had no meaningful existence and with more time to spare, wombats, sloths and bandicoots could probably have been found as well. It seems to be a characteristic of the human psyche that by thinking of topics in a somewhat obsessive way we can become highly sensitised to their presence. I have found numerous deserted medieval villages in this way and it works for other monuments. The danger is that by fixating on one interpretation of, say, earthworks a true identification may be missed: at one quite notable Yorkshire fortified site a deserted village was designated as garden terraces until very recently.

Landscape detection is an envigorating activity: it takes us outside and it stretches our minds. It would be dishonest to suggest that it is also easy – though in a world of dumbed-down standards tough mental exercise will do no harm. It is really like 'Cluedo' played on a vast open air stage, so that the countryside that has been evolving for thousands of years becomes the case and its myriad facets are the clues. The ongoing success in the UK of the *Time Team* series shows that there is plenty of public interest in such things – and perhaps the most worthwhile aspect of the programme is its introduction of the different fields of expertise that can be brought to bear on landscape interpretation.

Ripley

The Ripley study took place over a three-year period. It began almost accidentally when a survey of ancient pollarded trees in Nidderdale led to the realisation that some interesting earthworks were preserved in the deer park of Ripley Castle. In carrying bits and pieces of surveying gear to the park gate I became aware of earthworks in a field near the church, and this led to the investigation of a *lost* formal garden. Beyond the garden were fainter settlement earthworks, and beyond these, the site of the original church. Around this stage it was realised that the study could not be artificially curtailed and should span the whole township. The core of the work was concerned with the identification of earthwork evidence, while documentary study was usually able to relate these monuments to a historical context. Air photographs and old maps also played very useful roles. This was a project in non-invasive landscape archaeology: no excavations as such were involved. Some dating evidence was provided by pottery brought to the surface by rabbits or moles or found lying in the leaf litter on abandoned house platforms, while other materials were found eroding out of the graveyard at the original church site.

Introduction

Whether it is true or just patronising mythology that members of the Women's Institute would have competitions to see who could cram the most items into a matchbox, I do not know. The township of Ripley resembles such a matchbox: so many historic landscape features are packed within its boundaries that one may sometimes wonder if this is a real landscape or some gigantic archaeological teaching aid. There is a Roman road; an abandoned church; two deserted villages; a planned medieval village that is also a model village of the nineteenth century; a lost formal garden; a deer park, and very much more. Most of the features and monuments were unrecognised before the survey was undertaken.

A word needs to be said about how the book is structured. The end result of this complex process of detection is, one hopes, a story – the story of a landscape, and specifically here, a chronicle of the past times of Ripley. *Landscape Detective* is, therefore, organised chronologically, beginning with prehistory and taking the township through to the time of the Parliamentary Enclosures. Discussion of how evidence is found and interpreted is integrated into the text along the way.

This survey could never have taken place had either of the leading landowners, Sir Thomas Ingilby, Bt. or Kate Smith denied me access to their extensive holdings. In fact, unrestricted access was provided and I offer my sincere thanks. Ripley is reasonably well provided with public footpaths: the park is open to the public and worth every penny of the admission fee and I trust readers will appreciate that some of the places described are on private land. I would also like to extend thanks to Steve Atkinson for being so helpful in matters relating to access to different parts of the estate and also to the current rector at Ripley, the Rev. Stephen Brown. As for myself, despite a break of ten months as a result of injuries incurred while surveying at Ripley, I thoroughly enjoyed the work. Arriving at dawn on a winter morning *en route* to York to spend an hour searching for the faintest earthworks in the close-cropped pasture might not seem like fun, but it always proved to be the best part of the working day. Along with my editorship of LANDSCAPES, it provided a refreshing opportunity to engage in scholarly work.

Modest but essential grants for this project were provided by the North Yorkshire County Council Heritage Unit and the Harrogate Museums and Galleries Service. Assistance with archival research was provided by Dr Noël James Menuge and Matthew Holford. Help in the field was provided by the Ph.D. research students, Lisa Hammond, Joe Franklin, and Ian Dormor and the MA res student, Andrew Done. Ideas were exchanged with the archaeologists Steve Moorhouse and Kevin Cale. Ted Bishop provided interesting comments, while Richard Purslow of Windgather Press and my co-founder of LANDSCAPES has blessed the project with his enthusiastic encouragement. The original research presented here is the work of the author supported by the sponsors noted above. No research centre or institution of education has any claim to credit for this publication.

Reference

Muir, R., *Shell Guide to Reading the Landscape,* Michael Joseph, London, 1981.

To everything there is a season, and
a time to every purpose under the heaven:
A time to be born and a time to die;
a time to plant and a time to pluck
that which is planted;
A time to kill, and a time to heal;
a time to break down, and a time to build up;
A time to weep, and a time to laugh;
a time to mourn, and a time to dance;
A time to cast away stones, and a
time to gather stones together; a
time to embrace and a time to
refrain from embracing;
A time to get, and a time to lose; a
time to keep, and a time to cast away;
A time to rend, and a time to sew;
a time to keep silence, and a time to speak;
A time of love, and a time to hate; a
time of war and a time of peace.

Ecclesiastes 3: 1-8

CHAPTER ONE

A Time Far Away

We know very little about the development of the landscape at Ripley before the Roman period, so that by the time that we can begin, faintly, to perceive its outlines, the human-made countryside was already old. Before the Roman landings brought prehistory to an end in England it had passed through many stages. There were phases of expansion when the wildernesses contracted and phases of agricultural retreat, when fields and farmsteads at the margins of the working farmland would have been abandoned. Such cycles of colonisation and decay continued into the historical era, with times when society was stable and others marked by revolutionary change superimposed upon them.

Some information about the human impact upon the countryside can be gained from ancient pollen grains preserved in peat bogs on the Nidderdale Moors near the head of the Dale (Tinsley, 1975). In the New Stone Age or Neolithic period, birch woods crowned the highest ground, with mixed oak woodland on the slopes below and alder in the wet valley bottoms. Some temporary clearances were made, but it was not until the start of the Bronze Age, around 2000 BC, that a continuous assault on woodland and expansion of pasture was begun. By the closing centuries of the prehistoric era, the boundaries between pasture and moorland were similar to those of today. In the gentler valley environments the clearing of the wildwood will have been earlier and arable farming will have had a higher profile. However, ancient pollen sites are fewer and far harder to locate in the valley, so we must depend upon the evidence from the uplands.

Although the outlines of countryside were laid down in prehistoric times, in the township of Ripley there is no endowment of ancient monuments. In terms of the configuration of terrain, the most likely location is the knoll to the east of the village and just to the south of the road to Ripon known as Yarmer Head. Though probably without being one, it is reminiscent of a very small Iron Age hillfort. Thorpe, normally one of the more reliable Victorian historical writers, recorded that in 1830 some labourers, who were digging gravel to spread on the roads, broke into a brick vault set close to the roadside here, at a depth of six or seven feet (1866, p. 109). It contained two perfectly preserved skeletons which crumpled to dust on exposure to the atmosphere. In 1842 another skeleton was found nearby. It would only need a couple of subsequent writers to add some of their own speculations for a (false) prehistoric provenance to be created. Yet the skeletons would not have crumbled immediately on exposure, while brick-making was only introduced to Britain by the Romans, lost, and reintroduced during the medieval period. Yarmer Head was incorporated into a field pattern of pre- or early medieval date, so that subsequent vault-building would have been unlikely.

Thorpe was writing about a period 36 years earlier; he does not mention his sources and was probably drawing upon local hearsay. As a source for history, folklore tends to be unreliable and it is not the length but rather the shortness of popular recollections that can surprise. In this case, as in many others, there may be a grain of truth in the story, perhaps with the discovery of human remains providing the basis for the myth of the vault. If skeletons were discovered at the roadside and near the boundary of the parish then this might have been a typical Saxon or medieval execution site, with or without a gallows. Alternatively, in a time of crisis, the corpses of plague or battle victims might have been thrown into a convenient roadside gravel pit. Without the archaeological excavation of other skeletons, followed by a scientific forensic examination in a university laboratory, the facts will never be known.

The Romans landed on the Channel shores of (what would be) England in AD 43 and in the last third of the century they occupied northern England and invaded Scotland in the governorship of Julius Agricola in AD 78–80. Under Agricola, several fortified camps garrisoned by auxiliary troops were established at places such as Bainbridge, Ilkley and Bowes to assist in the pacification of the large territory of the Brigantes in the Pennines. Further east, commerce and administration assumed greater importance and new urban centres emerged. Aldborough, near Boroughbridge, was created as *Isurium Brigantum*, the administrative focus of this eastern section of Brigantium and probably named after the sacred river Isura, the Ure. With the establishment of such innovations, the essential character of Roman Britain emerged. There was an upland military zone, where the indigenous people were traditionalists and where emphasis was on policing and the extraction of raw materials, and a lowland civil zone, which was commercially active, receptive to Roman culture and technology, civilised and urbanised.

Connecting one of these zones with the other was the relatively minor Roman road linking the military camp at Ilkley (*Olicana*) to Aldborough, the walled town and administrative centre that covered around 55 acres (22 hectares) and lay on the road running from York north to Catterick and then beyond, to the defensive complex of Hadrian's Wall. This road spanned the military and the civil zones, linking the upland areas, where tribal customs remained intact, to the lowlands, where commerce was quickening and cultural horizons were broadening. Like most Roman roads, it was probably built at an early stage in the occupation to facilitate the movement of troops, but as the territory was bound more tightly into the Roman world, so economic factors will have come to the fore.

Researches with roads

Roman roads – or at least the missing or hypothetical ones – attract a remarkable amount of attention from amateur enthusiasts of various kinds. There must be something about them, perhaps the mental images of legions on the march, that strikes a popular chord in the imagination. Even so, ancient roads can be amongst the most unresponsive of subjects so far as landscape detection is concerned. To begin with, even if one succeeds in getting an answer right it is virtually impossible to prove it. Most of the questions are concerned with dating (is it a prehistoric trackway, a Roman route, a medieval thoroughfare, or what?). To validate an answer by archaeological means one would need to excavate a section of road and find datable artefacts – coins are easiest – beneath the roadbed and higher up, upon the old road surface. Then it could be claimed with confidence that the road was younger than the coin it had covered but older than the one dropped on its surface. In theory this may be fine, but in reality one cannot frantically go digging up sections of road. Even if one could, an awful lot of digging would be expected before any fortuitously lost coins would be turned up.

Landscape features can be dated relatively to each other. Thus, where a railway cuts obliquely across a system of fields it is plain that the fields were there before the railway. Equally, where a system of fields nests comfortably beside a road it is likely that the road was there first and when the fields were set out, they were adjusted to it. Generally, relative dating only offers very loose dates – like older/younger than a set of medieval fields.

Many searches are made for Roman roads but frequently they lead to disappointment. Not unusually, the trench that is dug to expose the predicted paved surface and roadbed reveals nothing of the kind. Belief in these roads can be rooted more in faith and imagination than in substance. Some of the problems concern knowing what minor Roman roads looked like, and appreciating that roads of similar appearance were built at a variety of times.

> The laying of a paved causeway a few feet in width was the only kind of road making recorded in medieval Nidderdale. Several causeways ... are mentioned in medieval and sixteenth century records, including the causeway of "Roudenscaha" [Rowden, near Hampsthwaite] and two causeways in Killinghall. The remains of one can also be seen in Potter Lane ... Some causeways were laid down for the first time in the seventeenth century, e.g. one on Nought Moor in Dacre (Jennings (ed.), 1983, p. 83).

Around 200,000 miles of paved roads served the Roman empire. Some of these were well over 60 feet (18.3 m) wide. Many on the continent were about 40 feet (12.2 m) from kerb to kerb, but most of those built in Britain were less than half of this in width. Some were built to a width of 22 Roman feet between the ditches, the Roman foot being about 11.65 inches (29.6 cm). In the north, (though seldom in the southern lowlands, where surfaces were of gravel or flints) roads were paved with slabs of stone. About as wide as an urban back street today, flanked by ditches and built of paving placed on a firm, well-drained base of stones and gravel, the Roman roads looked much like those built in the stony northern countrysides in medieval, Elizabethan and even Georgian times.

Various discoveries of 'Roman roads' may be proposed, but to merit serious consideration any candidate should form part of an alignment linking known Roman towns, ports or military bases. To appreciate how Roman routeways may appear today one may reflect upon the disintegration of the road network after the administrative hierarchy lost the ability to maintain the system. In places the drainage system would break down, allowing the roadbed to become saturated; feet, wheels and hoofs would churn-up the ruts and puddles until the road became a morass of mud, and potholes. Where this happened, travellers would abandon the damaged section, break into the adjacent fields or woods and establish a new routeway. In other places roads might be pillaged for useful kerb stones or paving stones, suffer intrusions by buildings, wells or cess pits or have their flanking ditches blocked by the builders of roadside dwellings or silted-up through neglect. In others still, the old alignment failed to link up newer settlements and religious centres, so diversions were developed. Much later, the construction of exclusive medieval deer parks or post-medieval landscape parks would close public access to sections of ancient routeway running through them. In all these ways, and others too, the integrity of Roman alignments was affected. Some routeways would seemingly vanish from the countryside, while more frequently, the routes would become hyphenated, with sections of the original alignment surviving in some places but being linked together by stretches of road developed to supersede 'lost' lengths of Roman road.

Generally, the lost roads do not vanish completely, though the nature of their survival varies. Some persist as holloways or troughs worn into the land surface by the traffic of centuries. Others endure as seemingly unremarkable lengths of lane, or have their courses traced by ancient hedgerows. Occasionally, portions of the old fabric may endure, though kerbstones are only likely to provide evidence where a number survive together *in situ.* Other evidence can be derived from place-names, of which the most useful are those involving the word 'street', which was used by the first English-speakers to denote a paved road. It should be remembered that during the centuries following the Roman withdrawal from Britain in AD 410, their roads remained as the most prominent features in the rural landscape. They could be employed in the compartmentalisation of the landscape into estates and townships and later, parishes. Old boundaries were frequently slung along sections of road, and may help to confirm the antiquity of such features.

The Ripley road

The road as it approaches, traverses and departs from Ripley is shown in Plate 1.

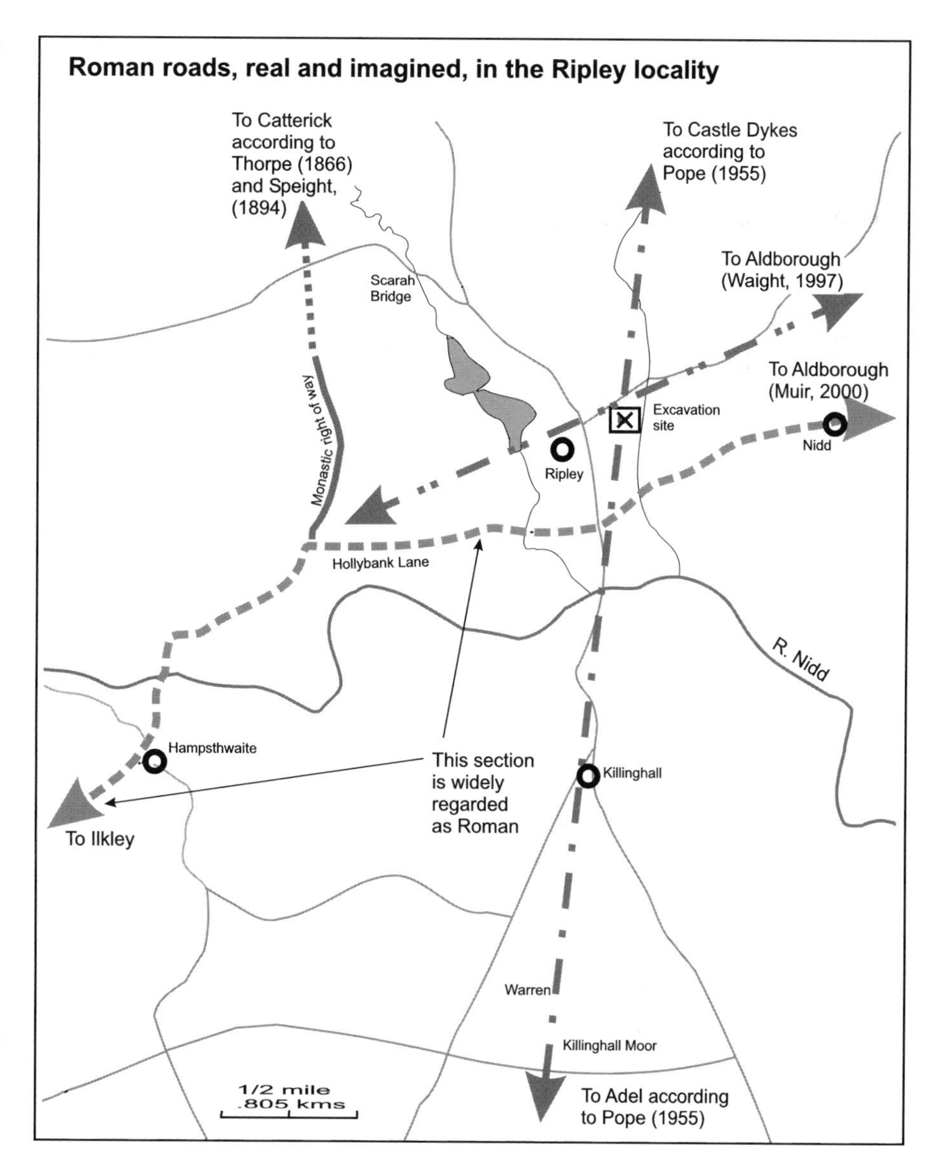

FIGURE 2.
Roman roads, real and imagined, in the Ripley locality. The map shows versions of the courses of various routes as proposed by different sources.

To those who imagine that Roman roads were uncompromisingly straight, it may seem unconvincing. Roman roads were certainly straighter than most. In the flat lowlands they did run straight, until obstacles like marshes or rivers were encountered. In the rougher terrain of the uplands, the roads ran as straight as was feasible. They tended to adopt direct courses across the plateaux, but would diverge where watercourses, steep gradients, marshes and other obstacles were encountered. In some cases the minor roads of the north seem to have adopted the courses of pre-existing trackways used by the indigenous people, and such tracks would have swung and curved. Possibly this was the case with the Ripley road, for the section where kerbstones remain to pinpoint the original course show a road that curved slightly in terrain that, though undulating, would have permitted the construction of an arrow-straight road.

Any readers thinking of applying their talents in detection to a locality should peruse the works of their local Victorian topographical authors: there was usually one and sometimes quite a few. As with Thorpe, mentioned earlier, they have to be read with caution. Archaeology was in its early infancy and versions of early history were clouded by the nationalist and invasionist obsessions of the nineteenth century. Once these writers begin to go wrong, their mistakes become springboards for deeper incursions into the absurd. On the other hand, they had distinct advantages. The worst assault on the heritage of countryside and monuments has taken place since the middle of the twentieth century, with the rate of destruction accelerating in the final third of the century. The old writers on history, topography and nature take us back to times when much of what has gone was then still in place, and much of what was in place had existed through many centuries of relative stability. They seldom reference their information clearly and freely borrow and pinch from, or embroider upon, each other's work. At the same time, they were much closer to the past than ourselves (Thorpe, p. 2) mentioned a native of Ripley who had died at the age of 104 and whose *grandfather* had held Cromwell's horse when the General appeared in the village. It is more credible because as a boy Thorpe had known this villager before he died, a little prior to 1820).

FIGURE 3. Kerbstones of the Roman road from Ilkley to Aldborough exposed in Hollybank Wood.

Some of the strengths and weaknesses of the Victorians are evident in this quotation from Speight (1894, p. 380) describing what he saw as he proceeded westwards towards the end of Hollybank Lane (Fig. 3): 'At the foot of the lane is Holly Bank Lodge [recently 'converted'], one of the entrances to Ripley Park, and beside it on the road is a small excavation from which in former times scouring-stones for the flagged floors of Ripley Castle were obtained.' Then he appears to make a mistake:

> On the further or west side of the Lodge is Back Lane, down which the Roman soldiery passed on their way from the great city of Aldborough to Ilkley. Remains of this old thoroughfare are still in evidence here in the shape of large pave-stones, some of which are as much as five or six feet long, and a foot in width. Many of them have been taken up for walling, &c., but many yet remain *in situ* along the line of Roman march, which can be traced over the river close to Hampsthwaite church ... (ibid)

Despite repeated claims by the old writers that Speight's 'Back Lane' (now a green footpath) was a Roman road it appears to be a medieval monastic right of way and heads for anywhere but Aldborough. It follows the park boundary southwards and joins Hollybank Lane just to the west of the lodge. On the other hand, Speight does provide some useful information. He explains the particular purpose of the little track-side quarry and, more importantly, he records that the roadbed had been robbed for walling materials. There is no evidence for massive kerbstones on 'Back Lane', which seems to have rubble beneath its turf, and it would be useful to know exactly from where the kerbstones were taken.

On the most elevated section of Hollybank Lane there is the best-preserved visible section of the routeway, a Roman designation being encouraged by the presence of massive kerbstones, which were used to hold the stones of the roadbed in place. With kerbstones recognisable on both sides of the road it is possible to measure its breadth, which is about 86 inches (218.4 cm) from outer edge to outer edge and 53 inches (134.6 cm) between the kerbs. This was plainly a very small road by Roman standards, but not so narrow as to prevent its use by laden carts. Further to the west in the wood, kerbstones can still be found, but some seem to have been disturbed and repositioned during repairs to the lane, while many more must have been robbed for other uses.

The nineteenth-century writers seem to have convinced each other that two Roman roads existed or met at Ripley, one being the convincing Ilkley/Aldborough routeway, the other being associated with Catterick. Grainge (1871, pp. 31–2) proposed a Catterick/Adel routeway that ran in an unspecified area somewhere near Ripley, crossed the river and continued southwards across Killinghall Moor. Thorpe (1866, p. 6) may have been the inspiration for Speight, for he, too, believed that a Roman road to Catterick left the Ilkley–Aldborough road at the western end of Hollybank Wood. He also mentioned the existence, across the Nidd, of 'Warren Camp', a proposed Roman encampment on the

FIGURE 4.
The Roman road heading towards the crossing on the Ripley Beck and thence to Aldborough.

old commons south of Killinghall (still an archaeological possibility). Much more recently, the story gained a new chapter of quite an amazing kind. If the story as reported in the *Yorkshire Evening Post* was true then in the late summer of 1955, a Miss Dorothy Greene was approaching Ripley while driving along the Ripon road towards Harrogate. She saw four old oaks aligned in a pasture and a slight ridge in the field, turned to her passenger, John Pope, a headmaster, and said: "Look, there is a Roman road". How anyone could identify a buried minor Roman road from a moving vehicle on such slight evidence is beyond my comprehension. Yet these were not charlatans or rank amateurs. Miss Greene was a fellow of the prestigious Societies of Antiquaries of England, while Mr Pope was a member of the Roman Antiquities Committee of the Yorkshire Archaeological Society.

The story did not end there. Mr Pope returned to the field during convalescence from an illness and on 29 September, 1955 he began to dig a trench (*memorandum placed with Yorkshire Archaeological Society*). As a result, he believed that he had found the Roman road overlain by a monastic road, and this was verified, with qualifications, by a professional archaeologist and a retired director of the Leeds City Museum. He dated the Roman road to the third century and thought that it could have linked the 'Romano-British' settlements at Castle Dykes near North Stainley to the little settlement at Adel near Tadcaster. Castle Dykes was, in fact, a Roman villa – and one cannot think of any compelling reason why it should have been given its own link to Adel. There are no fragments of routeway running north–south in the intervening countryside that might be the remains of such a road.

The monastic road, was said to have been built about 1190 under a charter granted to the monks of Fountains by Bernard of Ripley, giving them a right of way from their grange at Cayton to Ripley bridge (*Early Yorks. Charters*, I, p. 404). The conclusions based on this were mistaken. The right of way, 40 feet (12.2 m) in width '... *viam quadragnita pedum latitudinis* ...', did not run from a bridge on the Nidd near to the present Ripley-Killinghall bridge. Instead, it was actually away to the west and spanned not the river, but Ripley Beck to the west of the now-deserted village of Owlecotes at the site of Scarah Bridge: '... from the bridge of Rippelie Beck which is beside the house of Thomas of Ulecotes ...'. The track excavated by Mr. Pope could not possibly have been this right of way: he had misinterpreted the topography described in the charter.

The description of the supposed road as a ridge punctuated by four very old oak trees and the stump of a fifth immediately puts one in mind of an old hedgebank, of which only the bank and a few of the hedgerow standards or pollards remain. Although the map reference for the site provided by Mr Pope is inaccurate, the location concerned can be identified by his written description. He had, in fact, excavated the boundary earthwork of a group of plough strips or selions that comprised part of the scattered glebe lands that supported the priests of Ripley. (With better technique, he noted that the ridge underlay the hedges flanking the road to Ripon, this road having been superimposed across the fieldscape of plough ridges.) The most notable 'find' from the excavation was identified as a medieval 'cow-shoe' found beside the 'monastic right of way'. Such a shoe, shed by an ox from a plough team, might easily have been tossed on a nearby headland, joint or hedge bottom.

The 'lost' Roman road also intrigued a local councillor, E. C. Waight, who attempted, in 1963, to trace its course to Aldborough using documentary, archaeological and air photograph evidence (Waight, 1997). Well to the east of Ripley township, an excavation was undertaken at Waingates Farm. The line through Ripley proposed by Waight runs to the north of the one that I favour, it is uncompromisingly straight and, if accurate, it seems to have had scant effect on the subsequent development of the cultural landscape (Figure 2).

It is plain that intelligent and curious people can reach quite different conclusions about the existence and location of Roman roads. To enjoy a measure of success in this very demanding pursuit it is necessary to attempt (though this is never easy) to see things through the eyes of those long-dead people who made the important decisions. So far as Roman roads are concerned, the following principles seem vital:

A Without incurring ridiculous expenses or antagonising important people, the route should adopt as direct a line between A and B as possible.

B Where a chosen line can only be pursued at the cost of making major cuttings, causeways or drainage operations, or at the risk of exposing troops and travellers to ambush, then less direct alternatives should be considered.

C The terrain traversed should support and sustain a well-drained roadbed and be susceptible neither to prolonged waterlogging nor to sustained or violent inundation.

D In evaluating potential Roman roads the following points must be borne in mind:

1. The route should be aligned between two known Roman centres that one would expect to have been connected.
2. The route in question should be shown to be a part of a chain of Roman road 'fragments' sharing a common alignment and preserving the course of the routeway after the destruction of the intervening sections.
3. Finds in the form of goods lost, discarded or concealed by travellers will be associated with former roads and the distribution pattern formed by Roman finds in the area concerned should be considered.

In the case of the Ilkley–Aldborough road, there is a strong local tradition that regards it as Roman, with the 'known' section being known in the locality as 'Watling Street'. Also, Roman finds are associated with the zone through which it passes. Nine silver coins of the period AD 70–169 were found near Hampsthwaite in 1845, while Speight (p. 381) particularly mentioned bronze fibulae and female ornaments. Plainly, the road connects known Roman centres, while close examination of maps and terrain establishes the lanes, holloways and boundaries that trace the courses of the 'missing' sections.

The traditionally-recognised section of the road, through Hampsthwaite and beyond the River Nidd to Hollybank Lane is convincing. A cobbled surface 18 inches below ground level is believed to have been found at Hampsthwaite in 1927/28, while in 1998 a service trench was dug across the village, running through its green. No recognisable fabric from the Roman road was found, although the traces of a narrow but neatly-paved causeway across what would have been wet ground to the little Cockhill packhorse bridge just to the south of the village street and green were found. The track was considered to be medieval (K. J. Cale, personal communication). The lack of enduring Roman remains does not disprove the case for Hampsthwaite's Roman alignment. Over a century ago, Speight described the robbing of the road for building

FIGURE 5.
The Roman road discovered as an earthwork below the southern boundary of Ripley cricket field. The corner of a rectangular structure later diverted the track slightly to the right.

materials, and in a village with much digging and many demands for stone and rubble, the robbing-out of materials might be expected. The Roman engineers, in maintaining their line, had little choice but to ford the Nidd at or very close to Hampsthwaite and the village is plainly aligned upon an ancient route or trackway leading to a crossing on the Nidd.

Massive kerbstones are still in place on the loftiest section of Hollybank Lane. From the bottom of the lane, convention regards the road as bending and swinging through Ripley, passing between the castle and church and through the market square. This, I believe, is wrong. The track running into Ripley appears to be a medieval branch from the ancient routeway, either to the late-medieval village or to the manor which probably preceded the village.

Almost certainly a medieval mill dam on the Ripley Beck (located at or about the current crossing) was exploited as a makeshift bridge.

The Roman route has vanished into the fieldscape (and landfill) below the Lane. One factor governing the Roman choice of line that is not evident today was the presence of a huge meander loop that carried the Nidd much further north over its floodplain. The Nidd followed this course until it was straightened by the cutting of a new channel by Sir William Ingilby of Ripley Castle in 1665. In order to maintain a direct route towards Aldborough, the Roman engineers needed to cross the Ripley Beck quite close to the northern tip of the old loop on the Nidd. Their road must lie within a narrow zone – and in fact a short causeway across the floodplain of the Ripley Beck valley and the holloway of a route ascending the valley flanks were discovered by myself while mapping settlement traces in the pasture concerned.

The most difficult puzzle is found back in the Ilkley direction at the western end of Hollybank Lane (Plate 1). A Roman road should not bend sharply without a pressing reason. The reason for the sudden northern diversion near

FIGURE 6.
A massive old hedge runs beside the abandoned course of the Roman road in the east of the township.

the approach to Hollybank Wood may be found in the form of the old river bluff that lies just inside the wood and which would have presented its steep face to the traveller. It was not an insurmountable barrier, and a cutting could have been made to lower the gradient of any road that ran across it. However, there are no clear traces of cuttings or holloways in the wood: no convincing archaeological evidence of a former and more direct routeway. If the road was used by wheeled vehicles and heavily-laden packhorses (carrying lead from mines on Greenhow Hill above Pateley Bridge?) then the builders might have tolerated a more level but less direct route. On the other hand, the true track might still lie obscured and damaged beneath the wood and when great trees fall large craters are gouged as earth and stones are torn out with the roots.

The other puzzle lies to the east. After crossing Ripley Beck the road ascended to the plateau between the township's two becks (where it formed the boundary between the original open field furlongs or 'cultures'). It then descended to a crossing on Newton Beck. Today it is seen as a faint track beside an ancient hedgerow, though on a survey mapped by Chippendale in 1752 it still existed as a lane. Today, the hedgerow and track curve northwards towards a small bridge as they approach the valley bottom. However, a holloway can clearly be seen to leave their course and head directly for the beck. If the curving approach line to the stream represents the Roman road then its line would harmonise well with the grain of field boundaries. But if it trended straight across the little valley it is lost: obliterated in marl pits themselves obliterated by a sunken caravan park. Beyond this park and towards Nidd to the east it emerges prominently, with the 'agger' or raised roadbed flanked by ditches.

Documents as well as earthworks should be explored in any attempt to put a date on a routeway. The documents will not take the origins of the road back into the Roman era, but they may show that it has a long history and is not the result of post-medieval road-building.

Wood (1979) wrote that the paved road between Ripley and Clint was probably of the eighteenth century – though in fact its early provenance is good. Recording the ancient boundaries of the Forest of Knaresborough in 1613 Solomon Swale wrote: '... and so down the same lane by the same Pale to a Stone calld Cropp [Corp] Cross standing at Ripley Park side in the end of a lane leading from Clint to Ripley ...' Almost a century earlier, mention was made of '... the common highway through Ripley Park from Clint to Ripley Church and also to markets of Knaresborough, Ripon and Boroughbridge ...' (Ingilby MS 1065). The first cartographical representation of the routeway was printed in John Ogilby's atlas of England in 1675. Here the route from Knaresborough via Nidd and Ripley to Hampsthwaite is shown as part of an important routeway from York to Lancaster. It approaches Ripley from the east between 'lime pitts' and exits the village via the now mysterious 'Dark Hall' (plate 88 in Ogilby's atlas). Though part of an important routeway in 1675, previously the public status of the lane seems to have been debated, and in the 1650s the Ingilbys had clearly attempted to close the road leading through their park. A case was brought against their neighbour and the leading

FIGURE 7.
In its easternmost section the Roman road, its agger just beyond the fencing, forms the township and parish boundary between Ripley and Nidd.

landowner in the adjacent township of Clint, Thomas Beckwith, for digging up blocking posts placed in the road at the gate of the park at the western end of Hollybank Lane and entering the park with his carriage. A case of perjury was brought against one Miles Bolton for saying – quite truthfully – that the old lane was a common highway (Ingilby MS 3150).

The hunt for Roman roads exemplifies many of the pitfalls that threaten the landscape detective. Psychological factors should be borne in mind: once one becomes interested in a particular type of field monument one tends to see examples everywhere, with the brain becoming highly sensitised to the relevant evidence. Equally, once a particular explanation has been favoured it can tend to 'take root' with the exclusion of all contrary evidence and explanations. It is as well to develop a healthy scepticism where one's own interpretations are concerned and to ask oneself, from as objective a standpoint as possible: 'In this case, just how likely am I to be right in percentage terms?' With experience one discovers just how easy it is to be wrong in this most demanding of fields.

The same scepticism should be applied to the results of archaeological excavations. Though not present at the sites I cannot avoid a sneaking suspicion that either there are an awful lot of unsuspected Roman roads aligned on improbable destinations strewn about or that some of the 'Roman roads' excavated are, in fact, open field headlands or joints. As low ridges packed with cobbles and flags removed during land clearance and ploughing they are quite different from the fields around and at least superficially similar to the metalled roads of the Romans. These banks, aligned on the edges of fields, would gradually grow as successive generations of ploughmen scraped clinging clay from the coulter and ploughshare after ploughing a furrow. This earth would entomb the pebbles and stones cleared from the land and the headland would rise to a slightly arched section, like a cambered road.

In the case of my own identification of the Roman routeway, the main evidence came from the discovery of a holloway flanked by former hedgebanks and also a causeway by the crossing of the Ripley Beck. At the very end of this project, unexpected verification for this allignment came from the Ripley manor court rolls (Ingilby MS 1608, No. 8, 22 April 1631). In 1631 jurors were appointed to investigate a pattern of land ownership. They found

> ... lying at the south part of the orchard call *Parsons Orchard*, on the east part of the stream (*rivi*) there, and the south end of the same land abbuts on a parcel of land called *litle Chappell flatt*, and in the same way from the east part of the same; where it is shown to foresaid jurors that there was a way (*via*) in the foresaid land, and clear evidence (*plana mentio*) of a certain little raised way (*benelle cauminis*), and of a large ditch, in English the *cam* [i.e. ridge] *of an old* ditche, lying between the orchard and the foresaid land, heading towards the little stream (*rivulum*), with various dense thornbushes, in English a *young crabtree*, towards the east of the same.

This shows that almost 400 years ago the remains of the ancient road were more clearly defined than today, for I doubt that a modern jury of villagers could properly identify the earthworks.

References

Grainge, W., *The History and Topography of Harrogate and the Forest of Knaresborough*, John Russell Smith, London, 1871.

Jennings, B. (ed.), *A History of Nidderdale*, 2nd edn, Advertiser Press, Huddersfield, 1983.

Ogilby, J., *Britannia, Volume the First: or an Illustration of the Kingdom of England and Dominion of Wales*, 1675.

Speight, H., *Nidderdale and the Garden of the Nidd, a Yorkshire Rhineland*, Elliot Stock, London, 1894.

Thorpe, J., *Ripley: its History and Antiquities*, Whittaker, London, 1866.

Tinsley, H., 'The former woodland of the Nidderdale Moors (Yorkshire) and the role of early man in its decline', *Journal of Ecology*, 63, 1975, pp. 1–26.

Waight, E. C., unpublished memorandum: *The Roman Road from Hampsthwaite to Aldborough*

Wood, E. S., *Collins Field Guide to Archaeology in Britain*, Collins, London, 1979.

CHAPTER TWO

A Time to Mark the Outlines

The centuries following the Roman withdrawal from Britain around AD 410 are commonly known as the Dark Ages (though this term is frowned upon by many historians). In many senses the darkness was real and the period is regarded as a time of decay, instability and retreat from civilisation. All this is true, but it could also be a formative period during which the frameworks for medieval life and landscape were established. In the case of Ripley, many of the defining features of the countryside existing today seem to be traceable back to this period. The documentation for this period is so scarce that it is virtually prehistoric, yet it can still be brought to life through landscape archaeology.

Estate, road and fields

When the Roman road from Ilkley to Aldborough existed it will have been by far the most notable feature in the districts through which it passed. Though just a back road in a backwater of empire, it exposed the parochialism of this eastern fringe of the territory of the Brigantes to the cosmopolitan and civilising influences of the great continental empire. From time to time, cohorts of auxiliary troops, drawn from the corners of the Roman world and its barbarian periphery, will have been seen marching or trotting along the road. The Ripley locality was pacified and easily within reach of the garrisons at Ilkley and Catterick or the huge imperial fortress at York. Administratively, it was bound to the town and capital at Aldborough, while the commercial activities of the countryside must have quickened in response to the insatiable demand for food and produce from the towns and garrisons. Not surprisingly then, the road was a key facet of the cultural landscape during the Roman era, and it must have remained so after the empire withdrew. Roman roads did not vanish after AD 410. Most persisted and they continued to provide the fundamental English land transport network until the turnpike era of the eighteenth century.

My awareness of the significance of the Ilkley–Aldborough road dawned gradually. At an early stage in the research I became aware that the outlines of ancient rectangular fields were embedded in the fieldscape. They had to be old as some had had their corners detached by features like medieval roads – and such roads, though old, had yet to be younger than the fields that they

FIGURE 8.
Co-axial field system, probably laid out from the Roman road.

cut. Then I realised (as I perhaps should have done sooner) that the fields in Ripley – or at least the ones in the long-established areas of cultivation – shared a common north-north-west to south-south-east alignment. There are, in fact, the fairly well-preserved remains of a 'co-axial' system. These systems have been recognised in various parts of Britain and consist of large expanses of territory systematically partitioned to produce a gridwork of fields, all of

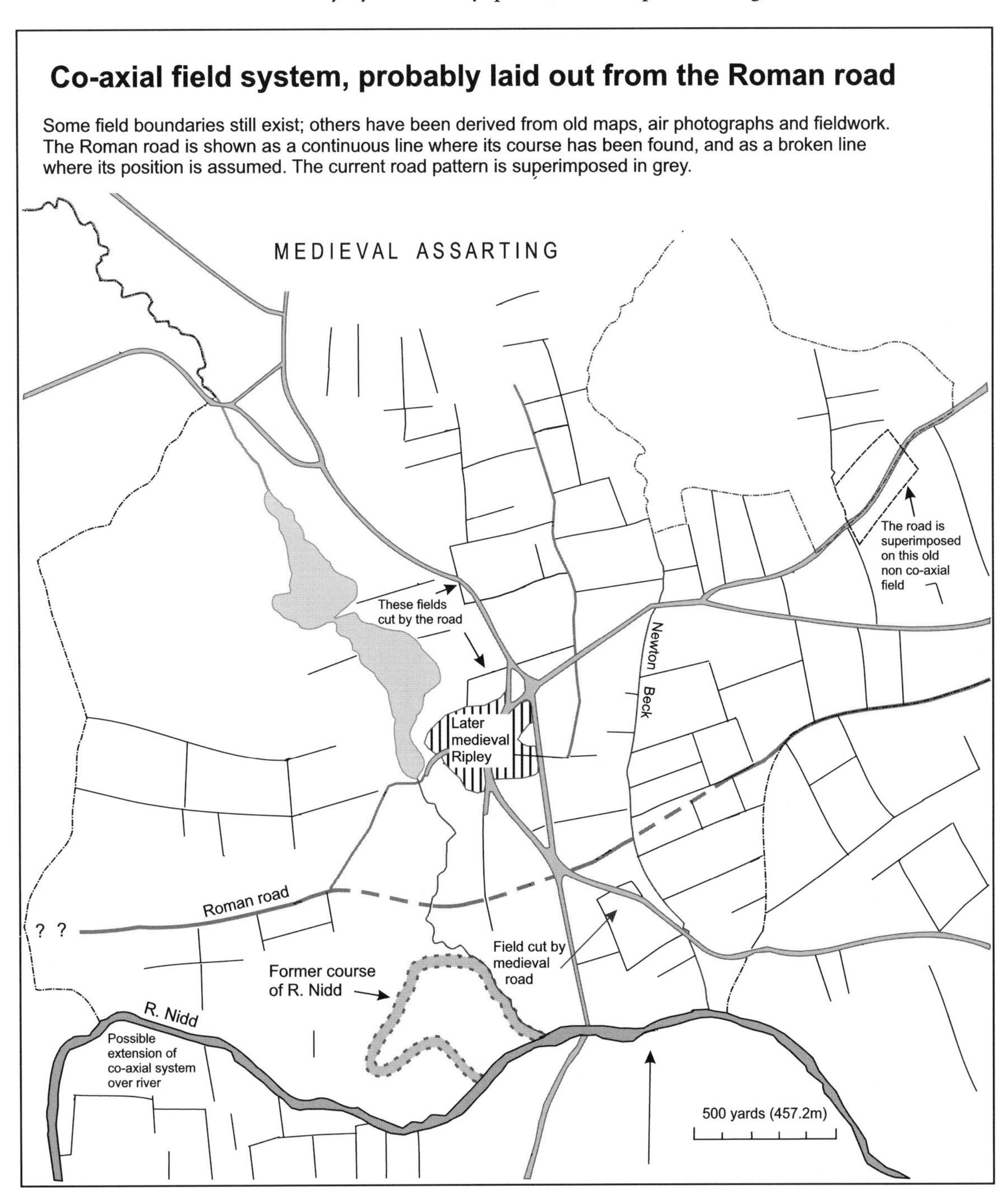

which are orientated on the same axis. Co-axial systems are not the distinctive creations of a particular phase in history or prehistory. The 'reaves' on Dartmoor have been shown to date from the Bronze Age (Fleming, 1988), the co-axial system on the Norfolk/Suffolk borders described by Williamson (1987) was cut by a Roman road and appeared to pre-date the Roman colonisation; geometrical field systems were occasionally set-out by the Romans themselves, the process being known as 'centuriation, while extensive prehistoric systems are numerous in the limestone Dales' (Fleming, 1998).

The field system in the Nidderdale-Wensleydale district survives in the areas which can be regarded as the agricultural heartlands of the region – those lands which would remain in cultivation longest during times of retreat and decline and which would constitute the historic core farmland of any vill. In various places it is possible to recognise (from the contrasting field outlines), a core of land with a NNW–SSE co-axial field pattern, and around it an arc of irregular enclosures produced by medieval assarting or land clearance. Beyond this zone of medieval colonisation are the geometrical fields produced by the eighteenth- or nineteenth-century Parliamentary Enclosure of commons. Elsewhere in the township, fields bounded by straight ditches that were produced by the drainage and enclosure of wetlands may be noted. Whether the co-axial systems were always confined to the agricultural heartlands or else became abandoned and obliterated elsewhere is not known, but to survive they must have gone into open field farming; have undergone enclosure, usually in Tudor times, and then have endured for a few more centuries. Given the time period and the nature of the changes involved, it will not be easy to recognise the dimensions of the original fields, but some furlong lay-outs seem to suggest that they developed from large fields, with sides sometimes measuring around 250–300 yards (228.6–274.3 m).

Open field farming will probably have been adopted in this area in the middle-to-late Saxon period – say around the ninth century. It has generally been considered that the introduction of an open field system – with its divisions and sub-divisions into fields, furlongs and strips or selions and its intricate operating arrangements – was a radical and traumatic event that swept away the preceding fieldscape. However it was possible for the elements of open field farming to be accommodated in a surviving prehistoric, Roman or post-Roman field pattern and: 'At Caxton, in west Cambridgeshire, the fragmented parts of a pre-existing, perhaps prehistoric, co-axial field system appear to have been fossilised within the open field furlongs and selions of its medieval landscape, some of which have survived into the modern period' (Oosthuizen, 1998).

Having recognised the existence of a co-axial system in Ripley township it became apparent that fields on the same axis were to be found across the river to the south, in Killinghall. This prompted a wider search, the results of which are shown in Figure 9, with the system that emerged spanning the area between the Pennine foothills and Boroughbridge.

Co-axial field systems could be created at any time, but they could only

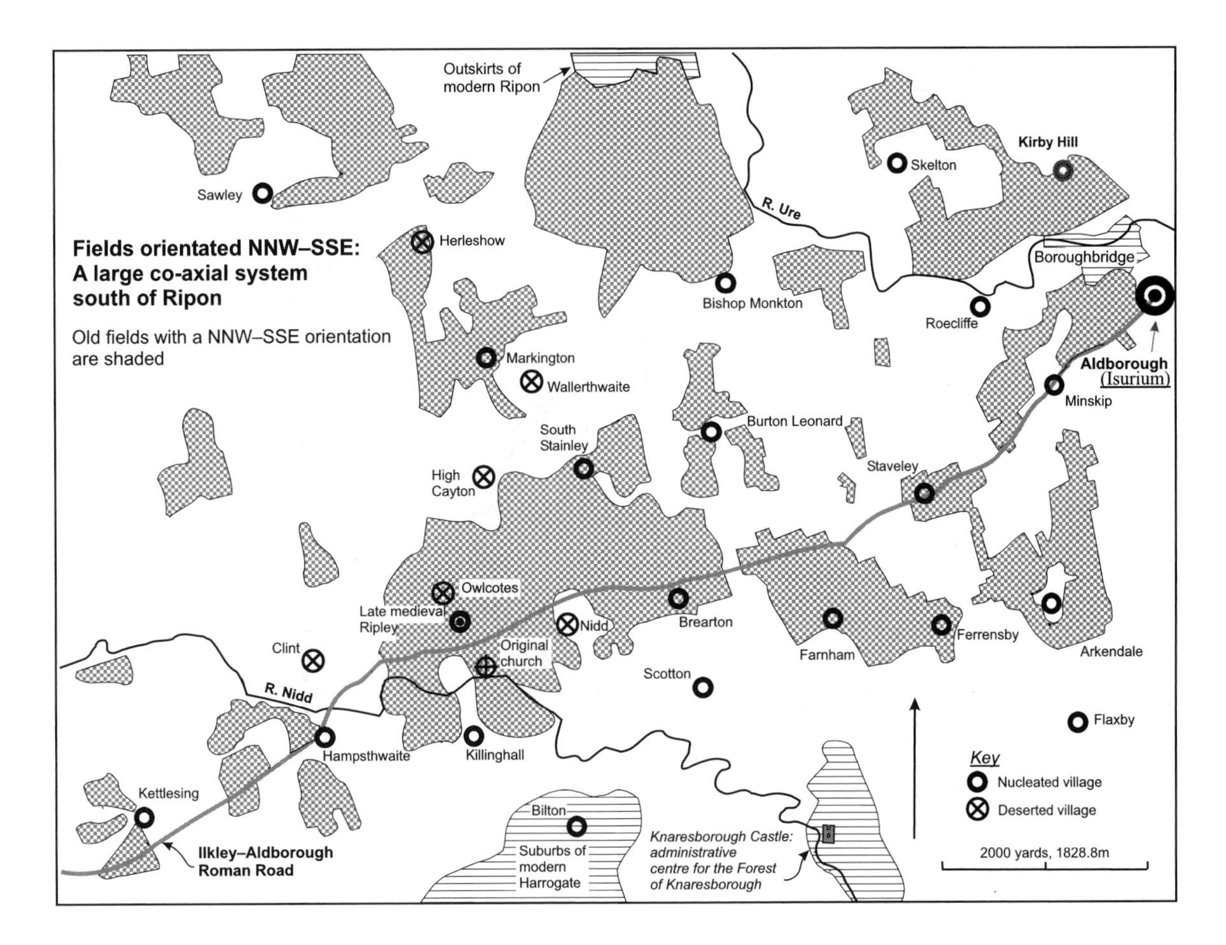

FIGURE 9. Fields orientated NNW–SSE: A large co-axial system south of Ripon. The areas with the co-axial fields seem to correspond with the local agricultural heartlands in which open-field farming would frequently develop. Relatively recent fields, like those derived from Parliamentary Enclosure, are not considered, although they may sometimes share the NNW–SSE orientation.

come into existence when special conditions prevailed. Some all-powerful authority, whether a chieftain, king, lord or council, had to be in control of an extensive territory. They had to be so firmly in control as to be impervious to the concerns of any tenants and communities who thought that their interests were threatened by the fundamental restructuring of their countryside settings. Co-axial systems could not be created today because of the detailed patchwork of the patterns of ownership, each occupier with his or her own priorities and interests to further, and each protected by bodies of accumulated legislation.

The Ripley co-axial system does not appear to be prehistoric because the field boundaries respect the Roman road. Indeed, the road seems to have provided the creators of the system with their fundamental alignment, the fields being laid-off at right angles to the great monument and axis. Thus, to date the co-axial system one has to find a period after the construction of the road and at a time when great expanses of land, the size of numerous parishes, could have been under a unified control. That time existed in the centuries immediately following the withdrawal of the Roman empire.

The earliest reasonably comprehensive description of the estate patterns of England that we have is Domesday Book of 1086. This does not depict the survival of an ancient system of lordship and land division, but rather the terminal stages in a prolonged process of fragmentation. Previously, territory seems to have been organised among much larger landholdings that have become known in modern times as 'multiple estates'. As described by Glanville Jones, the former Leeds University professor of geography, in 1971, multiple estates had their origins in the pre-Roman period and they had a number of components that might much later emerge as townships, parishes and smaller estates or farming units. Several of these components or their sub-units would specialise in certain fields of production that reflected the particular resources of their settings, like a horse-breeding complex or a cheese-making settlement. The lord of such an estate could visualise it as an integrated entity and organise its production to meet the needs both of self-sufficiency and commerce. Some of the multiple estates seem to have existed as late-prehistoric tribal territories, and many corresponded to miniature geographical regions, like river basins. In our case, the estate concerned seems to have had Nidderdale as its spine and to have extended to the Wharfe to the south and the Ure to the north-east.

This was the territory identified by Jones, as surviving in the form of the 'wapentake' (the northern equivalent of a hundred) of Burghshire that was mentioned in Domesday Book: '*Borgescire* was named after the Domesday *Burg*, now known as Aldborough ... This stands on the site of Isurium Brigantum, the fortified Roman cantonal capital of the Brigantes' (1979, p. 29). He went on to describe the boundaries of Burghshire:

> Neatly bisected by the upper reaches of the Nidd valley, the shire extends from a crest line in the Pennines down to the riverine barriers provided by the lower courses of the Ure, the Nidd and the Wharfe. Within this area there was a great variety of resources ranging from rough pastures and woodlands, especially at the higher altitudes, to large expanses of good arable lands in the lowlands (ibid).

Within Burghshire at the time of Domesday there were 135 'units of community' or vills, with a remarkably high proportion of these, some 60 in number, being entirely or wholly possessions of the king. In similar cases, the existence at the time of Domesday of considerable numbers of units or 'sokes' owing food and provisions to the king is regarded as a relic of an ancient system of tribute. The royal vills were concentrated in two royal estates. One was centred on Aldborough, reflecting a persistence of the role of a regional power centre inherited from the fortified Roman town of *Isurium*. The other *caput* or regional capital was Knaresborough. In both cases, the 'burg' or 'borough' element in their names denotes a fortified place; with Aldborough the reference is plainly to the old Roman defences, but the case of Knaresborough is more puzzling. The earliest reference to the castle is from 1129–30 (Pipe Roll 31, Henry I, p. 31), when there is a mention of 'the king's works', a term normally associated with castle-building, at the town, though different but unspecified defences here are suggested by Knaresborough's Domesday entry as *Chenaresburgh*. Both towns dominated rivers and their crossings.

Massively walled in red sandstone, Aldborough controlled the main Roman crossing on the Ure, while Knaresborough and its (later) Norman castle dominated the gorge where the river Nidd has gouged a deep channel through Magnesian Limestone strata and they were defended to the south and west by its cliff-like walls.

These two *caputs* appear to have controlled the remains of a vast estate represented by Burghshire, its origins perhaps receding into prehistory and its existence continuing into the time of English settlement. The estate became a possession of English kings, and during its fragmentation, its former status was indicated by the survival of a concentration of manors that were still royal possessions:

> It is ... possible that there was already a royal hunting forest in this area, the forerunner of the medieval Forest of Knaresborough, and that development had already been restricted in the interest of the chase. If such a hunting forest existed, it might help to explain the chain of manors dependent on Aldborough which stretched across lower Nidderdale into the Washburn valley, near the line of the old Roman road from Aldborough to Ribchester [and Ilkley], part of which still provided the main highway through the area in the twelfth century (Jennings, 1970, pp. 31–2).

It has been suggested that the two estates that can be identified from Domesday Book were originally just one unit, and that the Norman hunting reserve, the Forest of Knaresborough had originated as the woodland and hunting component of this estate. After the fragmentation of this estate: 'Components of both the main royal estates were intermixed alike in the area extending from Aldborough to the northern border of the forest, and within the forest near its southwestern boundary. This intermixture suggests that both estates had originally formed part of one unit' (Jones, 1979, p. 30).

Among the factors causing the fragmentation of the huge ancient estates were the granting of royal land to neighbouring estates (in this case like the one centred on the old monastic focus of Ripon), grants to favoured vassals and supporters (like the thane living at Killinghall) and ones made to the church. Another was the system of 'partible inheritance' that existed in the old Northumbrian kingdom, with lands being divided between the male heirs. By the time of the Norman Conquest, the nobles holding land in the Knaresborough region were the king, Merlesuan, Gamelbar and Gospatric. The indigenous nobility were involved in rebellions against the Conqueror and forfeited their holdings to the invaders William de Percy, Ralph Paganel, Gilbert Tison and Ernegis de Burun, though Gospatric repented and was allowed to retain part of his estates.

In Ripley – *Ripeleia* in Domesday – Merlesuan, and the old king's thanes, Archil and his son, Ramchil, had estates, with Merlesuan probably being the main landowner and the other two holding lands in Clint and Whipley and possibly owning what came to be known as Godwinscales to the south and west of Ripley Beck. Whipley may have contained a village of some kind at the time of the Conquest and is certainly a lost hamlet, while Clint village decayed during the eighteenth and nineteenth centuries (Muir, 1998). 'Merlesuan' may mean 'marl servant' (Thorpe, p. 9) and medieval Ripley did have extensive marl workings. To suggest that Merlesuan had an obligation to supply a superior from his Ripley marl workings would very probably be to press supposition much too far, for he was a major landowner with great estates in Yorkshire and Lincolnshire as well as lands in the south.

Many who are aware of no other historical sources will know of Domesday Book. The survey, commissioned by the Conqueror and compiled in 1086, is widely regarded as some sort of gazetteer of the medieval kingdom, thoughtfully set down to provide information for future generations of historians. It was really no such thing. It provided the king with an inventory of his tenants, dues and taxable assets, while rather than presenting a thoroughgoing description of the realm, the text is spare for the south and often so thin as scarcely to exist for the north. The section for Ripley is divided between its landowners, and, as translated by Lancaster (1918, p. 8), the entries read as follows:

Land of Ralph Pagenel.

Manor: in Ripeleia Merlesuan had four carucates* and a half for geld†. Land to three ploughs. Ralph has it and it is waste‡. In the time of King Edward, worth twenty shillings.

Land of the King's Thanes.

Two manors: in Ripeleie Ramechil and Archil [had] one carucate and a half for geld. Land to one plough. They still have it. In the time of King Edward, worth thirteen shillings. It is waste.

And in the recapitulation.

In Ripeleie: Ra. Pagenel four carucates and a half. In the same place the King one carucate and a half.

* A carucate, the equivalent of a hide, was notionally the amount of land that a family with one plough could plough in a year, generally amounting to four virgates or about 60 to 180 acres;
† A northern term for tax;
‡ Waste did not simply mean 'wasted' in the sense of 'destroyed' and it was also used to signify commons, both wooded and open. In the north it generally applied to lands ravaged during the Harrying of the North, but not necessarily entirely depopulated.

The Domesday entries tell us hardly anything about the landscape of Ripley. We learn the names of the landowners, but not the bounds of their holdings. We gather that the place had a substantial amount of ploughland, but that its value had fallen after it had been wasted. Anything else has to be based upon speculative deductions. There is nothing to tell us whether a village existed here, while the river, meadows, routeways, pastures, woods and church are among the other fundamental components of landscape that are disregarded. However, information from other sources does allow us to interpret the development of the countryside in the centuries leading up to the Norman Conquest.

The Saxon and Anglo-Danish countryside

To begin to understand the countrysides of the past one must discover how they were organised and partitioned. For example, the co-axial field system to the west of Aldborough could not have appeared, evolved and continued to be a visible element in the modern landscape without the existence of an estate large enough to incorporate it. Similarly, if we wish to explore the life and activities of rural societies, we need to reveal the geographical roots of community. We are brought up to regard parishes as the basic territorial building blocks of community. In southern England such an approach has a few merits, but in the north, it is wholly inappropriate. In the north, good land and the people to work it were more thinly spread. Consequently, churches were fewer and further apart and tended on the whole to be later in their origins. Parishes were large or vast and were far too big to express the grassroots character of a community. The Knaresborough region tended to be relatively well-endowed with churches; immediately upstream of the borough, these existed at Nidd, Ripley and Hampsthwaite. Further up the dale, the medieval population was thinner and dispersed and, until 1402 when a church was built at Thornthwaite near Darley, the next church after Hampsthwaite was not encountered until Pateley Bridge, about ten miles further up the dale. In the remoter places, churchgoing must have been an arduous undertaking more often neglected than observed by the scattered populations. Killinghall gained a church of its own in 1879, but the medieval parish of Ripley encompassed both Killinghall, to the south and Clint to the west.

The real building block of northern social geography is the township or vill, a land cell born out or the interactions between people and territory whose origins can seldom be identified and which must often burrow into prehistory. When the authors of medieval documents mentioned the vill of Ripley they were not denoting the parish, while much more often than not they were referring to the township rather than the village. Vills combined together to form parishes: 'The township seems to have been the smallest community division of those represented regularly in late Anglo-Saxon charters …' (Hooke, 1998, p. 70). Townships were, however, frequently themselves sub-divided to form little units of locality known, inappropriately, as 'hamlets'. In the case of Ripley, the hamlets of Godwinscales, Broxholme and Birthwaite might possibly be older than the township, though all of them might have resulted from episodes of medieval woodland clearance launched from the core of the township. Newton, the 'New farm settlement' has a name that implies a similar

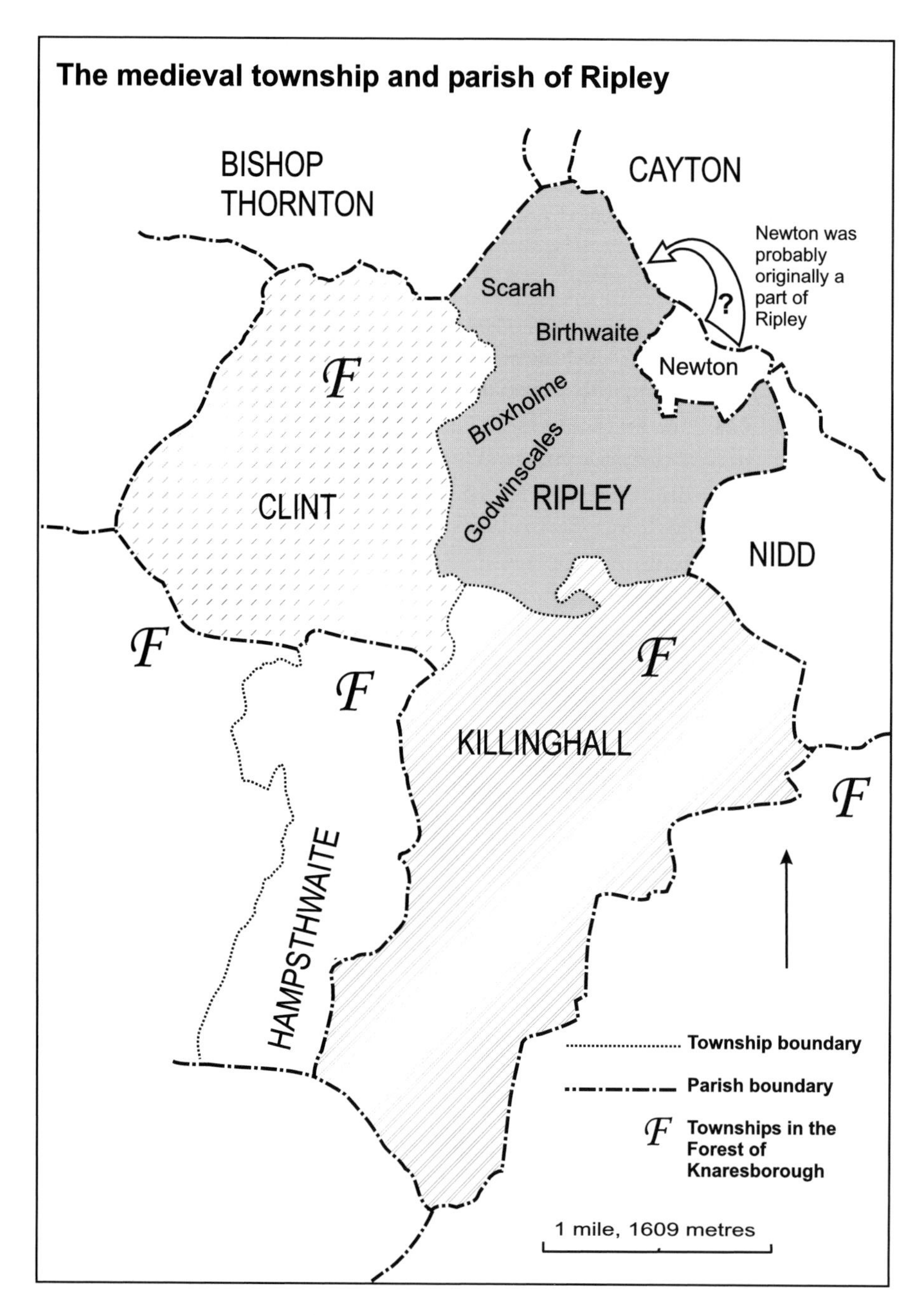

FIGURE 10.
The medieval township and parish of Ripley. The three townships in the parish of Ripley are shaded, while the Forest of Knaresborough included Clint and Killinghall townships, but not Ripley or Nidd. The geography suggests that Newton was a hamlet that was, for some long lost reason, detached from Ripley and awarded to Nidd. The boundary seems to have been adjusted to give Nidd a narrow neck of access to Newton.

origin, and this hamlet of Nidd township may once have been in Ripley township (S. Moorhouse, personal communication).

The boundaries of the township may help to reveal its history. Much the greater part of the boundary length of Ripley is associated with topographical features, like the river or Newton Beck or the Black Sike, implying a very old unit of land. The boundary with Clint is interesting. This boundary, adopted to mark the division between the Ripley manors and the royal Forest of

Knaresborough, is repeated in documents many times over. The strength and apparent antiquity of the boundary underlines the superficial nature of the medieval parish – for both Ripley and Clint were united as a parish while being separate as townships. Around 1200, in a grant of land from William son of Ketel de Scotton to Robert son of Huckeman, this boundary was described as: 'Beginning at the sike [ditch or streamlet] next to the bridge called Godwinebrigg [i.e. Scarah Bridge], and extending to the west as far as Godwinearch [Godwin's Oak], and thence to Lesebech, from one head [stream source] to the other, as the fee of Knaresborough is separated from the fee of Albini [in the way that the boundary separates the Forest of Knaresborough lands from those of Agatha Trusbut, wife of William de Albini], and from Lesebech towards the south according to the right bounds between the field [open field] of Clint and Godwinscales as far as the ancient fall [old course?] of the said stream to the river Nidd ...' (cited in Lancaster, p. 9). A few decades later, around 1240, the same land was granted by William de Goldesburg to the monks of Fountains and its boundary was described in identical fashion (Ingilby MS 176).

It was probably at the time when the lands in Godwinscales, by the western boundary of Ripley township, were in the possession of Fountains Abbey that crosses were erected where lanes entered these holdings. They appear in the boundary survey of the royal Forest drawn up in 1612

> ... and so up a little sike calld Black Sike running upon the outside of Ripley Park Wall or Pale unto [?Godwin's] Sike, and so as the said sike leadeth which runneth in the most place without [outside] the pale of Ripley Park and some places near the Pale within the Park to a cross calld Monk Cross standing in Whipley Lane End at Ripley Park yate [gate], and so down the same lane by the same Pale to a stone calld Cropp Cross standing at Ripley Park side in the end of the lane leading from Clint to Ripley, and so down by Robert [? Wood] by a little sike there to the waters of Nidd (Survey of the Forest of Knaresborough by Robert Wray).

One section of the township boundary does seem relatively young. In the south-east, between the place where the lane to Nidd crosses the Newton Beck and the Roman road the boundary is linked to a track which cuts obliquely across the strips and furlongs of the open fields. Here, the boundary must be younger than these features or otherwise the details of the fieldscape would surely have respected the boundary. The ridge and furrow is unlikely to be much earlier than the tenth century and, given the substantial amount of ploughland suggested by Domesday to have existed at the close of the Saxon era, unlikely to be much younger than the eleventh century. Possibly some readjustment of the boundary between Ripley and Nidd took place here in Norman or later times as part of an exchange of territory between leading landowners or when the land was under single ownership?

It is well known that place-names can convey information about past landscapes. However, quite a few theoretical edifices have been constructed upon the presumed evidence of place-names, only to crash down as new evidence emerges. The interpretation of place-names will never be an exact science because the exponents of the craft make choices that are partly subjective between different alternative translations of words. Thus, taking a Ripley example, the name Birthwaite might be translated as 'the clearing or meadow by the fort' by a historian deriving the name from the Old English *burh*, a fort and the Old Danish *thveit*, a clearing, meadow or paddock. More significantly, however, in a charter of the thirteenth century the place is 'Byrkenthwaite', showing that its meaning is really the (Old English) *birc* or birch tree -thwaite. In this way the name becomes useful, for it describes the nature of the woodland existing before people speaking old dialects of English and Danish cleared land for paddocks or meadows. It is also part of a cluster of names associated with woodland clearance which reveal the expansion of farming beyond its ancient core areas. In Nidd but bordering on Ripley is Hemming Syke Wood. Different translators might recognise the '–ing' component as either meadow (*eng*) or a group of followers/dependants (*ingas*) and the 'syke' as a ditch, but the 'Hemm-' might be derived from the Old English personal names such as Hemma, Helm or Hemede or from words denoting hemp, hops or broken land. Two or more possible translations exist for the majority of place-names.

The medieval charters contain a wealth of place-names, many of them associated with the small components of the farmed landscape. If these could all be linked to particular places then they would constitute a hugely informative archive of information about former countrysides. Unfortunately, most of these names and the things that they described perished centuries ago. Among the many around Ripley which cannot now be located are: *Midilundebrec*, the 'cleared ground in the middle grove'; *Bakstanerikes*, the 'place among the plough ridges where stones used for baking were obtained'; *Elineriding*, which might have been 'Ella's clearing', 'alder clearing' or 'flax clearing', and Thorphin's Croft. Others can be located, like *Bratheng*, or 'Broad Meadow', lying in the floor of the valley of Ripley Beck around the area inundated by the westernmost ornamental lake, or *Shandekefalde* – the 'cow fold belonging to someone with a name something like "Shand"' – which may have been a little to the west of the place where the Ripon road crosses the Newton Beck. A few names can be located but not translated with any certainty. Thus, Hencroft, mentioned in 1314, lay beside the then-still-existing village of Owlcotes, but though the name suggests a little piece of ground attached to a house and on which chickens were kept, the 'hen' -name could alternatively have described wild birds, stones, or a former owner called 'Hethin'.

Place-names can be significant when related to the old multiple estates. In this context, *-ham* names are sometimes associated with the leading settlement or main village of the estate, *-ton* or *-by* names with subordinate farmsteads, and names which describe resources and products, like sheep, cheese or barley, are thought to indicate the local specialisations in farming within the broad

FIGURE 11.
Ripley field names in 1807. The names recorded in a survey by Calvert and Bradley are of various dates and in many cases of debatable meanings. Dog Croft could record the kennels where hunting dogs were kept in the deer park, while Tanker Close may be developed from Tenter Close, a name recalling the medieval frames on which cloth was stretched. Great Saw Croft is by the river and could refer to sallow or willow, while Hell Kettles could describe hellish ground where charlock grew. Chapel Flatt is a furlong or culture named after the original but abandoned church.

estate. In our case, both centres are associated with the Old English *burh* names associated with fortified centres. Cayton and Newton border on Ripley, with Markington, Thornton and Brearton not far away. Brearton might be 'Briar Farm', but it is possible that they all have the *–ton* or farm element prefixed by the personal name of an owner. Both Ripley and nearby Whipley are names that appear to end in the Old English *leah* word, which, confusingly, can either describe a clearing or tell of the presence of trees. Whipley is not translated in full, and while the 'Rip' element may share a common meaning in Ripon and Ripley, I find it hard to accept the suggestion (Jennings, p. 27) that both were colonised by members of the Germanic Hrype tribe. Surely the tribe would have fragmented and its identity would have been lost in the course of migrating to England and engaging in the very slow northward advance of English-speaking peoples. Rather more convincing is the interpretation of Ripley as 'Strip-shaped Clearing' (Gelling, 1984, p. 205). Some of the

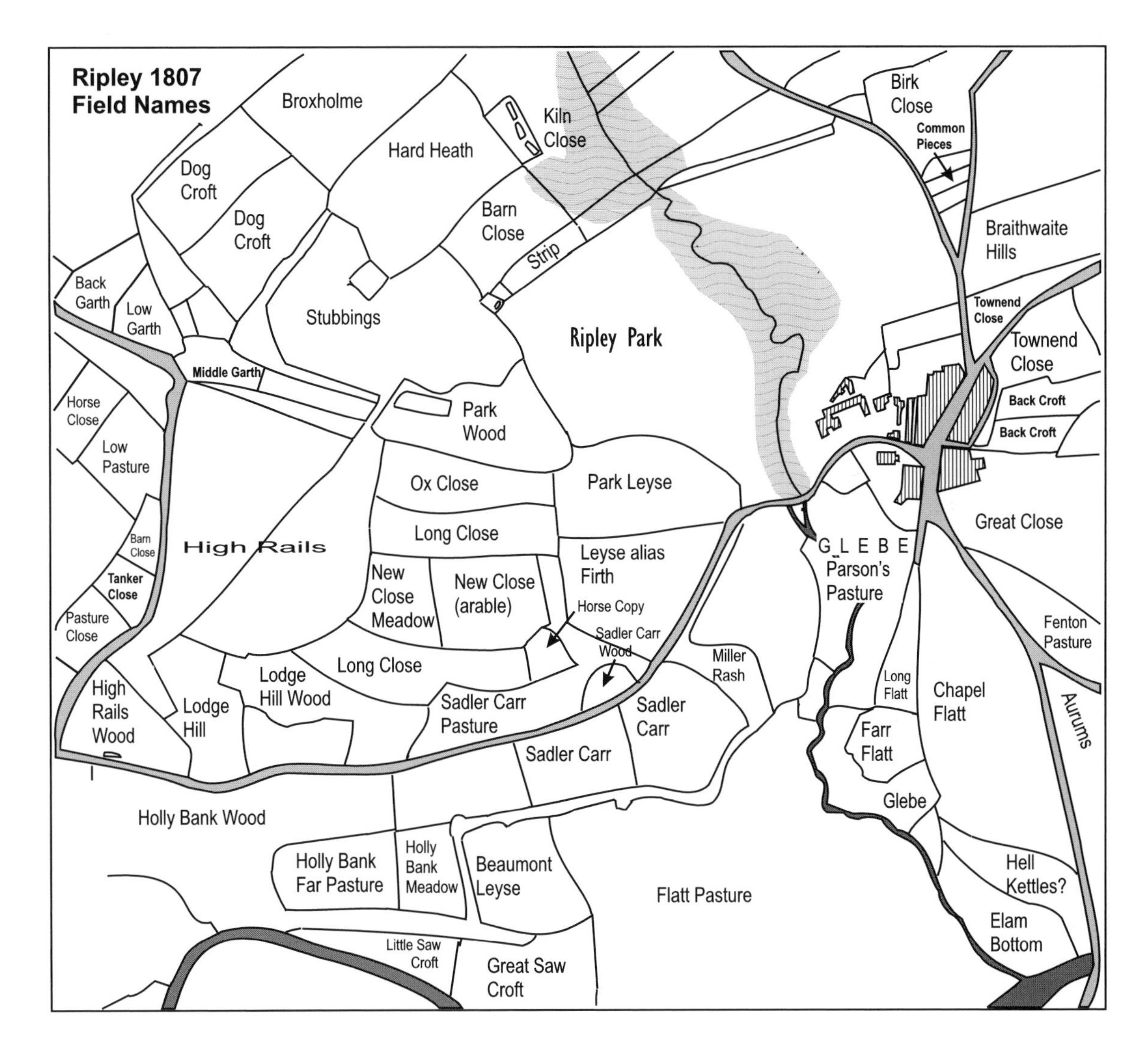

nearby place-names could reflect the division of labour on a great estate, with Swindon and Swincliffe both denoting wooded areas in which pigs were kept, while Staveley was the wood where staves were obtained. Such names certainly do not confirm the existence of a multiple estate, but they do throw light upon the nature of the pre-Conquest countryside.

Quite strikingly, the names demonstrate the character of Nidderdale and the adjacent areas under the English and Anglo-Danish kings as a cultural melting pot. English norms only gradually infiltrated and dominated the north of England. A Celtic kingdom that was probably Christian survived in the Elmet district in the south of Yorkshire until AD 617, when it was overrun by the forces of King Edwin of Northumbria, and Ceretic, the last of the British kings, was evicted. Less well authenticated is another Celtic kingdom of Craven in the Dales. The Celtic language survives in Nidd, the 'Brilliant or Shining River', perhaps in Dacre, the 'Trickling Stream', while Wallerthwaite near Markington might possibly have been home to an enduring community of 'Welsh strangers' or Celtic-speakers, (though equally the name could mean 'The Clearing of the Forest Folk' – or something else altogether). The victory in 617 must have been more a triumph for English culture and politics than a replacement of one nation by another. Before the English language gained an outright ascendancy in the north, the Scandinavians arrived. In the decades and centuries after the Viking raids on Lindisfarne and Jarrow, in 793 and 794 respectively, the Yorkshire Dales were affected by Danish settlement advancing from the east and by Norse migrants whose longships – or those of their forebears – had sailed from Orkney or travelled around the northern mainland of Scotland, shedding settlers on the western Isles, the Isle of Man, the Lake District, the Lancashire coast and that of Ireland. In Nidderdale the weakening Celtic traditions and those of their English successors fell under the powerful influences of two semi-distinct Scandinavian cultures. The surviving legacy of place-names tells of a remarkable time of cultural intermixture and adaptation.

Ripley's neighbours variously have the –ton and –ley names associated with English farmer colonists and the -thwaite and -scales of the Norse settlers, whilst Clint could be Danish and Nidd must have adopted the British river name. The cultural complexity was enduring and the names of people that were preserved in the earlier of the medieval documents relating to the manors show this. Ketel, Cnuth, Torphin and Godwin had Scandinavian names, while Edmund was English – though by the twelfth and thirteenth centuries, the William, Robert, Richard, Henry, Roger, Geoffrey and Ralph names favoured by the Norman aristocracy were beginning to be adopted by the lower ranks in the society of the dale.

The first church and village

In AD 597 Augustine's mission landed in Kent; in 627 King Edwin of Northumbria embraced Christianity in order to secure a marriage alliance with the Christian dynasty of Kent, and in 633 King Oswy of Northumbria presided at the synod held at Whitby where effectively the choice was made in favour of the Roman rather than the Celtic form of the religion. Religious issues apart, the conversion era of the seventh century was a formative and stressful period in the shaping of the landscape. Frameworks of ownership and organisation were being locked into place and the communities of literate monks were of great value to the kings. Minster churches were established on many royal estates, where the monks could record rents, boundaries and obligations and so provide the first accurate and authoritative record of the affairs of the estates. Later, during the last three centuries of Saxon rule, both the church and the lesser lay lords actively acquired estates and established their ownership under the framework of legal conventions.

There seems to be no surviving record of a pre-Conquest church at Ripley, but several factors related to the geographical setting do argue that there was one. The original church, the 'Sinking Chapel' (Plate 1), had a most unusual and difficult site, on a steep bluff overlooking what is now the former course of the River Nidd. During the winter of 1991–92, the Nidderdale-based archaeologist, Kevin Cale, undertook a rescue excavation after severe but avoidable damage to an adjacent medieval graveyard had been caused during work on a water pipeline. A fine item of Saxon copper work of the eighth century in a Celtic-influenced Lindisfarne style was recovered from topsoil adjacent to a burial. The nature of the piece is uncertain, but it might have adorned the binding of a Bible. Tweezers that might have been used for turning pages were also found. The discovery of the items of metalwork with probable religious associations provides at least the basis for arguing for the existence of a pre-Conquest church. There is also a more provocative possibility, for it has been suggested that this might have been the site of the synod of Nidd recorded by the Venerable Bede.

In 709, Bede recalled how Wilfrid had been evicted from the cathedral at Ripon and how he was rehabilitated:

> When they had studied the letters that Wilfrid had brought from the Apostolic Pope, Archbishop Bertwald and Ethelred very readily supported him. The latter, formerly king but now an abbot, sent for Coenred, whom he had appointed king in his own place, and asked him to be a friend to Wilfrid, to which he agreed. Aldfrid, King of the Northumbrians, still refused to see him; but he died shortly afterwards, and was succeeded by his son Osred. A synod was soon held near the river Nidd, and after some argument between the parties it was generally agreed that Wilfrid should be restored to the bishopric of his own church (Sherley-Price (ed.), 1968, p. 312).

Nidd has conventionally been regarded as the location of this synod, but the church there, of uncertain antiquity, is positioned at the edge of the parish, near the northern limit of the former village, and beside the old Roman road. It lies about one mile from the river as the crow flies. The first church at Ripley is beside the river, indeed it overlooks a little amphitheatre-like embayment of riverside lowlands which might have been ideal for the convention. Both church sites were well-served by land links, being on the Roman road in the case of Nidd and but a brief stroll away from it in that of Ripley, while both will have been near to any road heading up the dales from Ripon. In terms of water links involving the river, however, Ripley was much better served. The precise location of the synod of Nidd will probably never be found, but the Sinking Chapel site at Ripley does seem to mount the strongest single claim.

The medieval graveyard associated with the church lay on the level surface above the river bluff. (I believe that I have identified the remains of a ramp that linked the church to its graveyard, above). The rescue excavations showed that beneath the burials of the twelfth to fourteenth century that were destroyed by the contractor's trench there were some important features: post holes, three post trenches containing a number of post holes and a broad, deep ditch. The best interpretation of these features is that a sizeable complex of timber buildings – i.e. a village or conceivably a monastic settlement – occupied the later medieval cemetery site. Whether it was in existence as early as the synod of Nidd or whether it lasted as late as the Norman Conquest are not known.

FIGURE 12. The stone footings of the original church lying amongst the litter of the woodland floor.

Today, the situation of the Sinking Chapel, which is on private land, might best be described as 'appalling'. Trespassers who scale the barbed wire will encounter a forest of nettles in the summer, while fallen conifers, steep slopes and expanses of bog all guard the few remains of the church. When it was alive and functioning it was far from salubrious, the church being perched on a restricted shelf of river terrace at the foot of a very steep slope, while springs issued forth close to the ends of the building. Approaches, either from the east or the west, were difficult, and there are no traces of holloways or causeways to suggest that the church was reached from anywhere other than the village/churchyard area above. The ramp that I noted would have eased the ascent and descent, while the most practical approach may have reached the churchyard via a track running between furlongs that no longer exists. The slope behind the church must formerly have been stable, but has recently undergone rapid retreat, as evidenced by the undermined and exposed roots of trees growing at the top of the slope. I believe that during the currency of the church and village the earthen face of glacial sands, gravels and stones was revetted with river cobbles, concentrations of which can be found around its foot. The gradual disintegration of this cladding has exposed the slope to rapid erosion and retreat.

According to a survey undertaken by W. Chippendale, a local man related to the famous designer of furniture, the churchyard was trapezoidal in form, though almost triangular, with its broad base running along the top of the slope and its apex pointing roughly towards the north. Thorpe (p. 84) noted that an old wall, patched with fragments of gravestones, ran on the eastern side and he believed that the enclosure of the churchyard had formerly been surrounded by elms. He thought that several of their decaying trunks were still visible in the early decades of the nineteenth century. This would accord well with a much earlier account by Thomas Pennant, who made a tour from Alston Moor to Harrogate and Brimham Rocks in 1773:

> We ascended a small rising to *Ripley-Hall*, the ancient residence of the ancient family of the Inglebyes, at this time represented by Sir John Ingleby, a minor. The seat is at Ripley, a village placed on an advantageous bank, and well wooded: the guardian has lately spoiled it of its prime honours, some aged elms, which measured twelve feet six in girth, and were above seventy feet high (1804, p. 115).

(These trees could have bordered the other, surviving churchyard, where lines of trees are shown on Chippendale's map, while the old track from the village to its former church was said to be lined by elms in the eighteenth century.)

The Sinking Chapel plainly had a physically difficult and unattractive site. Space on the level terrace surface was limited, access was difficult and the importation of building materials must have posed considerable problems. The most striking features of this most singular of sites existed in the form of the springs. Sands, silts and gravels flushed-out of decaying ice masses form the superficial geology here. They are known locally as 'watter sands', which refers to the abundance of springs issuing from the deposits. Springs had a considerable magnetism where early church sites were concerned. This was mainly because, in the urge to commandeer the sites associated with pagan worship and acting on papal instructions, the missionaries would expropriate springs and holy wells from the pagan deities and then enjoy the glory reflected by their ancient sanctity. In this way springs or wells dedicated to the Celtic goddesses Annis or Elen would re-emerge under the patronage of Saint Anne or St Helen. Nobody would have contemplated building a church in such a place without a very good reason – especially with an abundance of firm, level ground lying just feet away. The springs provide an explanation for the peculiar situation of Ripley's first church.

Eventually, this site was abandoned. Local mythology claims that the church was undercut by the river and collapsed into the waters below. Another version holds that it was sapped by the flanking streams. In 1763–4 Edward Clough quoted current legends that ' ... the old Church stood upon the bank of the river Nidd, a little way out of the town, in a place (now) called the Sinking Chapel, as is handed down by tradition, because it sunk down into a quagmire'. Later he wrote that

> Thomas Ingilby is supposed to have rebuilded the Church at Ripley about the year 1400. The old Church stood a little out of town: and according to tradition fell into ruin by its critical situation on the Brow of a hill abounding with springs; the place continues to this day, to be prey'd upon by the springs in the same manner as formerly, without a possibility of stoping [sic] it, tho' the surface for the 10 or 12 feet thick is firm and good (Ingilby MS 3757)

Speight (1894, p. 353) provided one of the more colourful accounts:

> There can, however, be no doubt that the church stood on the elevated tongue of land called Chapel Flat at some distance from the union of the two streams, and that the undermining action and gradual widening of the waters at this point, aided by frequent floods, led to the destruction of the protruding land and consequent collapse of the church. The grave-yard adjoining the building seems to have partly subsided too, and fragments of bone and portions of coffin-wood have been, I am told, often found in the disintegrated gravel and broken sides of the bank.

It is true that the flow of water was greater during the medieval period, for the river Nidd had yet to be diverted and its waters greatly augmented those of its tributary, the Ripley Beck, which now has the old channel to itself. Even so, I doubt that the river undermined the church and even in flood the Nidd was probably too far from the wall base to occasion real harm. The actual cause of the abandonment of the old church is described in Chapter 4. Rather more seems to have remained of the original building when Thorpe was writing in 1866; he wrote of finding: '... the foundation of an old wall, perhaps some eighty yards or upwards in length ...' These cathedral-like proportions must be an exaggeration. Today, all that can be seen of the building is a single broken course of foundation stones traceable for about 20 yards (18 metres) and aligned both along the terrace and in the conventional east-west orientation. If these represent the last of the *north* wall of the church, then the building has been lost to archaeology. If, as seems likely, they are the remains of the *south* wall then the floor of the building may be partly *in situ* and intact. The old church was pillaged for most of its stone as well as its rude screen as its successor was built.

The actual antiquity of the original church at Ripley is unknown. Thorpe (pp. 86–7) wrote that:

> Whether the foundation of this sacred structure is to be ascribed to the Saxons or early Normans cannot be satisfactorily decided, for we are

> told that the old *Rood Screen*, now in existence, bore the date of 1040, which if correct, must belong to the former period; yet strange to say, it is not named in the Domesday survey. The earliest valuation of the living we have met is that of Pope Nicholas IV, AD 1292, when it is stated to be worth £23 6*s*. 8*d*. per annum. In the 'Nova Texata', AD 1318, the value is only £10. This diminution was owing to the devastation of the Scots ...

We can take the date of the church back a little further than Thorpe's 1292, for a Bernard the clerk or priest of Ripley was recorded as the grantor of land close to the church site to St Peter's Hospital, York in 1190–1210 (Chartul. of St Leonards; Rawl. B455, f. 91d.; *Early Yorks.*, vol. I, p. 84). Perhaps a little earlier, in 1170–1200, Bernard was mentioned again in connection with a grant of land, a mill and the right to appoint a priest to the church at Ripley (Ingilby MS 167).

Only an excavation could establish the antiquity of the church. Sometimes, the creation of a diagram can help to clarify the drift of the evidence:

Factors Arguing for pre-Conquest Origin	*Factors Arguing for a post-Conquest Foundation*
1. The association with (pagan holy) springs	1. The absence of a mention in Domesday Book
2. The discovery very nearby of a fragment of ecclesiastical metalwork presumed to have come from a bible of the eighth century	2. The absence of recognisable pre-Conquest masonry or statuary
3. The close proximity to a settlement, probably secular but possibly monastic, which most probably dates from the late-Saxon period	3. The apparent lack of evidence of minster status.
4. The possible association with the Synod of Nidd	

One could increase the precision of such a table by adding weightings to each entry. Each of the first three entries in the left-hand column is important, so they could each have a notional weighting of 3, while the lack of Domesday evidence does not seem to merit any positive weighting as this survey has little or nothing to say about most of Nidderdale, other than to record past and present owners and the values of estates. Structures from earlier churches in timber or stone could easily have been destroyed in building the last stone church to stand on the site. (The rood screen in the surviving church at Ripley was apparently moved from the 'Sinking Chapel' and cut down by two feet (0.6 m) to fit the new church; it is reputedly from the reign of King John (1199–1216) but stylistically it appears to belong to the fourteenth century.) The first Ripley church need not have been a minster, the role of mother church being played by the renowned foundation at Ripon, just a few miles away.

A search amongst debris eroding out of the churchyard produced parts of a roof tile and a floor tile. Together with the rood screen, these suggest that the church that was abandoned had undergone a rebuild within a century of its desertion and was a well-appointed building, roofed and floored in tile and provided with a fine oak rood screen in the fourteenth century. Several cinders from a furnace were found, perhaps suggesting an industrial site, but quite likely being associated with equipping the church, as for example, with the on-site casting of bells or metalwork.

FIGURE 13. Building platforms appear to lie upon level steps in the staircase of terraces on the north bank of the former course of the River Nidd. Tangles of elder and alder, toppled conifers and ubiquitous springs make this difficult ground to cross today, but formerly the south-facing riverside terraces would have been attractive places for settlement.

The former church is not the only building to have left its traces on the stepped river terraces which face southwards across the old meander loop of the Nidd. On both sides of the church site are the traces of rectangular platforms which appear to have carried buildings. They lie within the alder wood and elder scrub, with the exception of one small platform that overlooks the Ripley Beck, just to the south of the causeway and holloway of the Roman road. These platforms vary considerably in size but are roughly rectangular in form and are situated on level steps in the river terrace system (see Plate 4). Concentrations of river cobbles suggest that the scarps or risers fronting these platforms were stabilised by a revetment of stones. A curious feature concerns the 'hollowed paths' that appear to lead from the (presumed) entrances directly down to the river. As these approach the platforms they become more deeply incised, suggesting that the buildings may have had large roofs of thatch that swept down well below head height, thus favouring lowered entranceways.

FIGURE 14.
The landscape historian, Steve Moorhouse, standing on one of the building platforms near to the original church at Ripley.

FIGURE 15.
Now deeply covered in leaf litter, this platform is typical of the presumed early settlement sites.

The dating of this settlement is problematical. Most platforms are covered in a deep accumulation of peat or leaf mould and no surface traces of pottery were found, suggesting that the community might have been aceramic (lacking in pottery). South-facing river terraces such as these will have been favoured at all periods in prehistory, and Mesolithic flints are present, but the rectangular form of the buildings argues for a Roman or post-Roman date. This might be an extension of the (?) late-Saxon settlement that occupied the plateau above, but the impression is not of an agricultural settlement, but rather of a community that turned its back to the land and its face to the river, which must have offered opportunities in fishing and trade. In medieval times the head of navigation was downstream at Nun Monkton, where the Nidd joins the Ouse and where the water-borne goods were transferred to pack horses or carts (Beresford and St Joseph, 1979, p. 10). By using lighter craft and portages to by-pass the shallows and rapids, a greater flexibility could be enjoyed.

The settlement also presents problems because the house platforms, instead of being cut into or raised upon a sloping or uneven natural surface, are set upon natural river terrace platforms. Three other working archaeologists have looked at the site, and all were convinced by the platforms, so perhaps my slight reservations reflect an excess of caution. Only an excavation could determine the dates of the settlement. It lacks the road and lane structure one associates with a medieval village, and I would be a little surprised by a post-Conquest date. It might be contemporary with the original church, or with the pagan use of the holy springs.

Interestingly, the grant made by Bernard, the priest of Ripley, to St Peter's Hospital in York in 1190–1210 (*Early Yorks. Charters*, I, p. 84) tells of: '... tres acras terre de cultura nostra in Erburghouet juxta Nihd' (three acres of land in our culture or furlong by the earthwork near the Nidd). This might also be translated as 'Three acres at the head of, or above, the earthwork ...' (S. Moorhouse, personal communication). The area of the medieval township that was bounded by the Nidd was limited and the lost 'Earthworks' location must lie in a narrow zone that corresponds to the Sinking Chapel location or to the lower course of the Newton Beck. Around 1200, the original chapel was intact and functioning and certainly not an earthwork. The best candidates concern the adjacent settlement traces or else marl workings possibly already active in the limestone just east of the Newton Beck – but such workings would presumably have been referred to as marl pits rather than as earthworks. The traces of a settlement that was abandoned and yet which remained sufficiently prominent to serve as a local landmark might appear to be the best identification for the lost 'Earthwork' location.

The centuries from the withdrawal of the Roman forces in AD 410 to the Norman Conquest truly constitute an age of darkness. Earthworks, documents and facets of the living countryside provide some clues to the period – sufficient to offer glimpses or else the opportunity to reach false conclusions. The picture created here is one in which the Roman road was employed by the organisers of a great, Aldborough-based estate as a convenient alignment from which a co-axial system of fields was superimposed across huge areas of working countryside. With the establishment of Christianity, three churches were founded in the middle section of Nidderdale, at Nidd, Hampsthwaite and Ripley, each one either beside or very close to the ancient highway. The Ripley church is likely to have been an early foundation, associated with the appropriation of pagan holy springs. In late Saxon times, a substantial village developed on the level ground just above the church, while another settlement, orientated towards the river, existed on the nearby river terraces perhaps before the church and perhaps contemporaneously with it. Its remains appear still to have been a significant landmark in Norman times. Some important features of the present landscape existed by the time of the Conquest, but others had still to form.

References

Beresford, M. W. and St Joseph, J. K., *Medieval England, An Aerial Survey*, Cambridge University Press, Cambridge, 1979.

Fleming, A., *The Dartmoor Reaves*, Batsford, London, 1988.

Fleming, A., *Swaledale, Valley of the Wild River*, Edinburgh University Press, Edinburgh, 1998.

Gelling, M., *Place-names in the Landscape*, Dent, London, 1984.

Hooke, D., *The Landscape of Anglo-Saxon England*, Leicester University Press, London, 1998.

Jennings, B., *A History of Harrogate and Knaresborough*, Huddersfield, 1970.

Jones, G. R. J., 'The multiple estate as a model framework for tracing early stages in the evolution of rural settlement' in Dussart, F. (ed.), *L'Habitat et les paysages ruraux d'Europe*, Université de Liège, Liège, 1971.

Jones, G. R. J., 'Multiple estates and early settlement' in P. H. Sawyer (ed.), *English Medieval Settlement*, Edward Arnold, London, 1979.

Lancaster, W. T., *The Early History of Ripley and the Ingilby Family*, John Whitehead, Leeds and London, 1918.

Muir, R., 'The Villages of Nidderdale, *Landscape History*, 20, 1998, pp. 65–82.

Oosthuizen, S., 'Prehistoric Fields into Medieval Furlongs? Evidence from Caxton, Cambridgeshire', *Proceedings of the Cambridge Antiquarian Society*, 86, 1998, pp. 145–52.

Pennant, T., *A Tour* [in 1773] *From Alston-Moor to Harrowgate and Brimham Crags*, John Scott, London, 1804.

Sherley-Price, L. (ed.), *Bede: A History of the English Church and People*, Penguin, Harmondsworth, 1968.

Speight, H., *Nidderdale and the Garden of the Nidd, a Yorkshire Rhineland*, Elliot Stock, London, 1894.

Thorpe, J., *Ripley: Its History and its Antiquities*, Thomas Hollins, Harrogate, Whittaker & Co., London, 1866.

Williamson, T., 'Early co-axial field systems on the East Anglian boulder clay', *Proceedings of the Prehistoric Society*, 53, 1987, pp. 419–31.

CHAPTER THREE

A Time for Growing

In the old school history books the period known as the Middle Ages is often shown as running from the Norman Conquest to the Dissolution of the Monasteries. In more refined versions, a watershed is often located in the middle of the fourteenth century, when the devastation caused by the Pestilence of Black Death transformed the social and economic patterns. In Ripley the years after 1350 witnessed remarkable changes, although the effects of the Black Death were only a part of the story – and one which left relatively little lasting impact on the landscape. The preceding centuries were a time of growing, a time when the recovery from the retreat and decay that had followed the Roman withdrawal from Britain accelerated, when agriculture advanced from the old farming heartlands and when new farmsteads and hamlets abounded. The dawn of the period was far less hopeful, and was marked by a systematic campaign of terror that was waged against the people of the North.

The Harrying and the monks

Early in the fourteenth century, Scottish war bands burned Knaresborough and terrorised its surroundings. Some violence may have accompanied the initial colonisation of Nidderdale by Norse and Danish Vikings. None of this could have compared with the Norman Harrying of the North in 1069–71. In 1068 the northern earls, Edwin and Morcar, had withdrawn their support from the Norman monarchy, which appears to have been particularly unpopular in the Anglo-Danish North. Two successive Norman earls of Northumbria were assassinated, and the English claimant, Edgar Atheling, was proclaimed king in York in 1069. The Conqueror recaptured the city, though in the autumn of that year a large force of foreign raiders led by Sweyn of Denmark was welcomed in the northern capital. William rode north again, this time with the intention of finding a permanent solution to his northern problem. This took the form of the Harrying, seemingly a systematically organised campaign of slaughter and destruction – one that would leave the countryside still crippled and partly empty of people a decade and a half after the events.

Recently, authorities have questioned the severity of the Harrying and its legacy (e.g. Palliser, 1993 and Austin, 1990). Certainly, some of the claims relating to the size of the Norman host, the level of casualties and extent of devastation that it caused may, like many medieval statistics, be of a rhetorical nature. Even so, there is no doubting the fact that chroniclers living in an age when injustice and cruelty were commonplace were shocked by the scale and evil nature of the vengeance inflicted upon the North. Orderic Vitalis, who was born four years after the Harrying and who must have grown up amongst people who remembered its horrors, wrote of the terrible famine which followed the killing and burning and caused 100,000 Christian men, women and children to die of starvation. He implied that an unhealthy predisposition towards sadism tainted the Conqueror's character: 'Nowhere else had William shown so much cruelty. Shamefully he succumbed to this vice, for he made no effort to restrain his fury and punished the innocent with the guilty' (Chibnall, 1969, p. 231). Simeon of Durham also described the terrible aftermath of the Harrying, while the wasting was reported in the Anglo-Saxon Chronicle. Given the modesty of the documentation that survives from Norman England, the Harrying is a well-recorded event and it seems harder to doubt its awfulness than to credit it.

The plainest testimony of the devastation is found in Domesday Book. The suppression of the northern uprisings revitalised the process of stripping estates from their indigenous lords and allocating them to continental supporters of King William. Of the 1,782 vills in Yorkshire, 480 were recorded as still being totally waste in 1086, and 314 were partly waste (Darby and Maxwell, 1962). Within Yorkshire the pattern of devastation varied, with the Norman host apparently devastating areas along the main routeways along which it travelled and making forays into parts of the Dales in pursuit of resistance fighters (Bishop, 1962, pp. 1–10). Nidderdale above Knaresborough was particularly severely affected, with all the territory that became the Forest lying devastated apart for the low values accorded to Bilton and Whipley. Ripley, where the two estates had been valued at £1 and 13*s*. for the period just before the Conquest in the reign of Edward the Confessor, was simply recorded as waste. The apparent lack of people or economic activity deprives us of information about the division of classes within the local social hierarchy or about the presence of 'specialists', like priests or millers.

Whether the devastation and depopulation was as complete as one might imagine is another matter. Waste could indicate useful commons as well as abandoned farmland, while it is possible that numbers of local people survived the Harrying, but were then shipped-out in attempts to re-populate more valuable estates in the lowlands. In the case of Ripley's Norman owners, the King obviously had a multitude of other estates, while Ralph Pagenel was a rich and important figure who became Sheriff of Yorkshire under William Rufus. He owned Leeds and Headingley, as well as numerous estates in Yorkshire and beyond, ones that he had taken-over from Merlesuan, the native lord.

Wasted land had its own particular attractions. At the time of Domesday Book the Dales will have been quite extensively wooded, while farming in all but the lower sections of the Dales was of a marginal and poorly-productive nature. The Harrying, which had been particularly severe in some parts of the region, had removed populations which might otherwise have been an impediment to the chase. It had, in short, created ideal hunting territory. The king's Forest of Knaresborough, the first surviving mention of which dates from 1167, followed the Nidd for many miles and inclusion within its bounds imposed measures and obligations for the protection and management of deer upon the inhabitants. Desolate country also appealed to the Cistercians, founded at Citeaux in Burgundy in 1097 in an attempt to create a purer, more ascetic monastic order. In 1132, a group of dissidents broke away from the large Bendictince house of St Mary's in York and settled beside the river Skell just a few miles from Ripley. They gained enrolment into the Cistercian order and built their abbey of St Mary *ad fontes* – Fountains Abbey – which soon gained, by a variety of means and practices, a huge agricultural and industrial empire which eventually stretched into the Lake District.

Ripley narrowly escaped being engulfed in one or other of the great domains. The Forest followed the southern bank of the river but, perhaps because of some history of royal ownership, looped across the Nidd to include the township of Clint. Therefore, Ripley had Forest territory both to the south

FIGURE 16.
The present landscape of the neighbouring township of Cayton, which fell completely under the control of the Cistercian monks of Fountains Abbey.

and the west and its parish lay partly in the Forest of Knaresborough. In 1135–57 the monks were granted two carucates or ploughlands (probably amounting to over 100 acres, 40.47 hectares) in Cayton, just across the township boundary (Chartul. of Fount., f. 254). Fountains Abbey also developed an interest in Ripley and came close to controlling the township. In the thirteenth century it secured a strong presence in the west of the township in the district known as Godwinscales, roughly equating to the castle park. In the years before 1250, Thomas of the leading de Ripley family gave the monks of Fountains all his land in Godwinscales, and the monks acquired the remaining Godwinscales land that had been mortgaged to the Jews of York by the king's sergeant William de Goldesburg (Ingilby MS 173). More territory was gained in the northern part of the township, adjacent to the abbey's large holding in Cayton. In the event, Fountains failed to gain control of Ripley and towards the end of the medieval period the abbey's holdings were sold to lay landowners. Had Ripley been taken into the royal hunting Forest the consequences might had been a slowing of medieval land clearance but they would not have been severe. Had Fountains gained outright control of the township, however, the situation for the local people could have been extremely unenviable. The monks acquired the adjacent township of Cayton with the result that peasants were evicted, their settlement and tillage were extinguished and all were replaced by a system of farming organised from a grange and accomplished by lay brothers.

The years of monastic ownership in Ripley and Cayton have left a legacy of monuments, most notably in the forms of crosses and fishponds. Crosses were commonplace in the medieval era and served a variety of purposes. In general they can be likened to an 'X' placed on a map: they announced that there was something significant about the spot concerned, and they might also seek to hallow it. The place in question might be a burial ground, the limits of sanctuary offered by a great church or some other significant boundary. Ripley was well-stocked with crosses, and although nothing written remains to explain their functions, they do appear to have been put in place to mark the bounds of the land owned by the monks.

A cross stood beside the side of the deserted village of Owlcotes at the junction of the roads from Pateley Bridge and Masham (Thorpe, 1866, pp. 118–19). This is borne-out by the field name 'Cross Close' shown on the 1838 Tithe Map in this general vicinity but a little further east. A cross in the eastern extremities of the township at Yarmer Head is mentioned in 1414 (Ingilby MS 206), and this would have been close to the spot where the boundaries of Ripley and Nidd townships met that of Fountains' possessions at Cayton. A cross known as Monk Cross must certainly have marked the western limits of the abbey's possessions in Godwinscales and Robert Wray's survey of the Forest of Knaresborough of 1612 mentioned '... a cross called Monk Cross standing in Whipley Lane End at Ripley Park Yate [gate] ...' It still stands today, but the building of a park wall in 1813 has left it stranded and it is now surrounded by conifers. Further down the boundary to the

FIGURE 17.
The last of the crosses marking the boundary of monastic territory in Ripley to remain *in situ*. It lies in a coniferous plantation standing by the green of the lost hamlet or small village of Whipley.

south, another cross, known variously as 'Cropp', 'Corps' or 'Cap' Cross stood at the point where the bridleway leading from Clint to Ripley crossed the old Godwinscales/Forest of Knaresborough boundary. This cross has vanished – or has it? A re-erected cross which stood on the side of the green at the village of Clint, which was gradually deserted during the eighteenth and nineteenth centuries, looks very much like a small market cross. But there is no record of a market at Clint, where one would have been too close to and have been out-classed by the markets at Hampsthwaite and Ripley. Could the Clint cross be the old Cropp cross, transported just a quarter of a mile or so and set up on the green as an act of whimsy by post-medieval owners like members of the Swales family who had squabbled with the Ingilbys about the closing of the adjacent Hollybank Lane? I think that it probably is.

Medieval fishponds or 'stews' could include sets of breeding ponds, but generally they existed as larders in which wild fish caught in adjacent water bodies could be stored alive. Stews were particularly valued by monastic communities. Not only did they provide a reliable supply of protein that was available throughout the year and which could be exploited whenever parties of guests appeared, but their fish also substituted for the meat that was denied by religious conventions on certain days. Some stews were artificial containers scooped in the ground or embanked with boulders and clay and placed in valleys above the limits of flooding. Others were created by ponding back the waters of a small stream behind a dam. Where the stream concerned was also a boundary water, the consent of owners on both sides was needed, for the

FIGURE 18.
Oaks growing on the great dam on the beck between Cayton and Ripley that was cast up to pond back the waters to form a fishpond.

impounded waters will inundate valley bottom lands in both ownerships. Thus, about 1200, Bernard the priest of Ripley and his brother, Richard, granted the monks of Fountains '... the attachment of their stew on the donor's land to the height of 12 feet near the meadow of Nicholas de Cayton, namely from Godwin's clearing up to their other stew, and so far as water of that stew covered the donor's land in Ripley (Chartul. of Fount.; Tib. C. xii, f. 256). Shortly before, Richard de Ripley had granted the monks the 'attachment of a stew' upon the 'Water of Dalbeck' (Cayton Beck). The fishponds formed a significant part of the economy of the grange. The massive medieval earthwork of the fishpond dam (at grid reference SE 28596276 and presumably the one sanctioned by Bernard and Richard) survives as one of the best examples of its kind. Nearby, the traces of fish curing houses survive on the Ripley side of the stream in Cayton Gill Wood, and on the eastern flank of the stream the slipways used for boats are recognisable (S. Moorhouse, pers. comm.).

Monks, fishponds, boundaries and lost crosses combine in a colourful and not always reliable history of the Ingilby family written by one Edward Clough in 1763–4. Of the land provided for the monks to make their pond on he wrote:

> By the original grant the covenant of Fountains had liberty to enclose this piece of land with hedge and ditch ... And do whatever they pleased with it; so that the fence now does of right belong to the owners or occupiers thereof and not to Ripley ... It was marked by crosses cut on

> stone at the bottom of the hill; two of these still remain close to the fence wall on Scaro moor, viz ... in the angle made by the Horse pasture fence and the other about 80 yards nearer Cayton Hall (p. 37).

Despite the apparent readiness with which the local landowners donated territory to the monks and provided them with rights of way, the relationship was not always friendly. In 1269 there was a dispute between Reginald, the Abbot of Fountains and William de Ripley and his men. The Ripley party attacked brother Henry Wither, the granger at Cayton, on the common pasture in Ripley and they impounded his cattle. The issues were solved by holding a loveday and by William conceding that cattle from Cayton were entitled to graze on the Ripley common and by paying a substantial fine. Fountains Abbey was a very powerful interest, and not one that a fairly minor member of the local gentry could lightly upset.

The countryside

Despite the horrors at the dawning of the era, the Norman period became a time of rapid expansion that was sustained until the start of the fourteenth century. The climate was friendly and encouraged colonists to farm ground that had frequently been considered to lie beyond the frontiers of agriculture. Population was also expanding – though probably not to the levels that had existed in Roman times. Areas of woodland and thorn scrub that had spread across the Roman fields were hacked back and new farmsteads and hamlets were established to house the swelling population. The pattern of peasant cultivation that existed at Ripley was echoed in hundreds of other townships. The good, low-lying and well-drained land was designated as ploughland, while the damp, flood-prone ground fringing the river and becks was common meadow. Pasture land was limited, but available on the common or waste, on some cleared ground and in the open ploughlands, both on fallowing land and in the stubble of the cultivated furlongs after the harvest was taken. The expanding population put pressure on the farming resources of the township, with assarts being made into the woods and wooded pastures of slopes leading up to the waste.

Fields

The common ploughland was organised according to the complex patterns of open field farming, with great fields being divided into blocks, known locally as 'flatts', in the north generally as 'cultures' and more widely as 'furlongs'. The furlongs, which would have been accommodated in the field divisions of the old co-axial network, were divided into 'selions' or 'lands', now commonly known as 'strips'. These were very roughly an acre in extent and took the form of ribbons of land, sweeping slightly in the shape of a reversed 'S' or reversed 'C' when seen in plan. This curving outline was produced as the long plough team of six or eight oxen was swung into the turn as the headland at the end of the selion was approached. The selions were the basic units of land tenancy in the arable area, with each family working a collection of them that was scattered across the furlongs of the township. Each one existed as one or more plough ridges, the land being deliberately corrugated by alternating ridges and furrows, perhaps to assist the drainage and perhaps to secure the survival of at least a part of the crop either in damp furrows or dry ridge tops during seasons of extreme climate. At Ripley, the lands of the two manors seem to have been interdigitated in places, with selions scattered together.

The documents are surprisingly silent concerning the basic open field pattern, though the township will probably have had at least two and possibly three or four fields. In 1495 the Abbot of Fountains was granted marl from land formerly in the abbey's possession that lay '... in the south field of Ripley' (Y.A.S. MS 284). On the other hand, in 1538, when the common ploughlands were well on the way to extinction, land in virtually the same location was described as lying in the east rather than the south field (Ingilby MS 1065). If the field divisions were not considered important, the system of crop rotation could have been operated on a furlong-by-furlong basis. Some of the medieval furlong names are preserved in the documents. Thus in 1414 a feoffment or deed of gift from Thomas Vasavour to Thomas Ingilby exchanged two acres in Hallstedes and a quarter of an acre and a quarter of a rood in Fowlow in the demesne [the land farmed for the lord] of Ripley for one acre and one rood in Langlandes, a furlong that the agreement tells us stretched from two boundary stones to Newton Beck, and one and a half roods on Wyndegates, a furlong on the south side of the Ripon road, which was beside the cross that then stood at Yarmer Head (Ingilby MS 206). The 'Langlandes' furlong name must refer to the unusual length of its selions or lands, while a rood was usually a square measurement of a quarter of an acre.

Mentions of the land divisions can be found for earlier medieval periods, though as one moves back in time it becomes progressively more difficult to identify the places linked to the field names. A feoffment from Roger de Methelton to Alan son of Nicholas de Cayton concerned a quarter of an acre (a small selion) in the northern part of the division known as Torphin's Croft, a quarter of an acre at Handeran, one rood of meadow towards Whaiteholm, meadow land at Hallestat and as much use of the township's common as was attached to these lands, while Alan could also put ten pigs in the township wood each year, if he so wished. Whaiteholm could have been the holm or meadow near the thwaite or clearing, while Hallestat seems to have been by the manor house – but otherwise the locations of the places are mysterious.

In Ripley, plough ridges of various forms can be found, and they can also be seen in various stages of destruction by later ploughing. The medieval examples dating from before the fourteenth century are broad and where well-preserved – as with the set of ridges that approaches (and underlies) the gardens and back yards at the northern end of the village from the west – they have pronounced corrugations. In a few places, relative dating is possible,

with the narrow ridges (difficult to detect from the ground) that overlay the former parson's garden below the rectory, dating from the eighteenth century. In the Park, ridges that are both narrow and straight result from the use of a plough that was hauled across the land on a cable by steam engines in the nineteenth century. Later steam ploughing, still organised to produce ridge and furrow, is among those factors which can make the interpretation of ridge and furrow earthworks a challenging exercise. In addition, in the Park sets of plough ridges will suddenly be seen to terminate without any headland being apparent. This is the result of wartime ploughing during the twentieth century when rectangular patches in traditionally uncultivated areas were brought under the plough in response to crises caused by German attacks on convoys. When immersed in the problems of medieval archaeology one tends to lose sight of the transformations wrought in the modern era – even although this is the greatest era of change and destruction that has ever been known.

In many parts of rural England it is still possible to reconstruct most or much of the medieval arable landscape from the fragments that remain and/or

FIGURE 19.
Near the centre of the photograph a rectangular enclosure (allotments) is superimposed on ridge and furrow that has been cut by the road to Ripon.
The modern village lies immediately beyond the upper margin of the picture.

FIGURE 20.
The ridge and furrow in the centre of the picture has been cut by late-medieval roads running from the planned medieval village. The ridges run under the village.

were recorded on maps and other documents. The following techniques of landscape detection may be applied

1. Where ridge and furrow survives, either strongly or faintly, its orientation and form (reversed-'S', reversed-'C' etc.) should be plotted on a large scale map. Look out for the most frequently-adopted medieval standard in the area concerned and pace-out this dimension to assist the detection of ridges in the most difficult fields. In Ripley this measurement is around five yards (4.5 metres).
2. Using the oldest reliable field map available, note those fields that have curving boundaries. These are likely to pre-date the Parliamentary Enclosure of the eighteenth and nineteenth centuries. Look particularly for those boundaries that have reversed-'S' or reversed-'C' outlines, as these fields will almost certainly have been formed from the amalgamation of adjacent selions. Shade the former arable area accordingly.

3. The more difficult exercise of filling in the gaps in the medieval arable map must now be tackled. Consult the oldest available air photographs for any subsequently-destroyed traces of ridge and furrow and scour the photographs and the real fieldscape for any low ridges or swells which could be former headlands. Using information from current farmers and old maps and surveys, attempt to identify any field names that are characteristic of ridge and furrow cultivation. Rigg = ploughridge; land = selion; Flatt, culture = furlong, head = headland (sometimes) are examples. Consult the medieval documents and find which of the fields and furlongs mentioned there are capable of being identified today. It is possible that a document may allow the orientation of ploughridges to be deduced. Thus, a document of 1276 mentions '... le heuedland at the bankeclose in Eum ...'. It shows that a headland followed the boundary between the field called Eum or Elam and the enclosure on the (river)bank. Therefore the selions in Elam must have run at right angles to the river and headland (Ingilby MS 188).
4. When the reconstruction map is as complete as it can be, examine the boundary patterns. Try to discover – so far as one may – whether the medieval pattern obliterated the preceding fieldscape or whether an earlier system, perhaps part of a co-axial network, was adopted to serve as a framework for open field farming.

Low-lying land beside watercourses was plainly designated as meadow land. Although such land would at first have been held, like the arable, in dispersed strips or 'doles', the absence of earthworks produced by ploughing meant that this crucial land-use left few traces in the landscape. Only chance discoveries of the stones used to mark the division into doles may commemorate the age-old cycles of mowing, drying and storing hay. In Nidderdale the most common meadow names are those ending in –ing, -eng and –holme. Ripley was typical in having its main hay meadows or 'engs' on the damp ground fringing its watercourses. Deliberate or accidental flooding by the two becks in early spring would have fertilised the land with silt and stimulated the grass into early growth.

Within the open fields, and sometimes even after enclosure, stones seem to have been often used as boundary markers. Moving them might involve attempted thefts of land and would in any event cause serious inconvenience, thus in 1649 one John Flesher was fined when he moved 'le bounder stone' between the lands of 'lord' Vavasour (of Newton) and Scarah Moor, the common (Ingilby MS 1607, no. 13, 16 April 1649). Flesher plainly had some point to prove for he had been fined for the same crime a year previously (Ingilby MS 1607, no. 19, 16 October 1648). The common itself was the subject of a variety of offences, which generally involved the unlawful privatisation of common resources. A James Skelton was doubtless perpetuating a medieval practice when he was fined in 1631 for digging three cart loads of turf from the common (Ingilby MS 1607, no. 12, October 1631). In 1621, William Steele was fined for taking a cart load of sand from Scarah Moor (Ingilby MS 1607, no. 26, 17 October 18 Js. I).

Woods, wood pastures and hedgerows

In the medieval lowland township the arable cultivation held the centre of the stage. It commandeered the best land and demanded the heaviest investments of time and labour by the countryfolk. Its complicated arrangements for the pooling of beasts to form six- or eight-ox plough teams and for the management of each niche in the tilled area produced the carbohydrates needed to support a swelling population. Other forms of land-use, in their ways no less crucial for survival, had to be fitted around the ploughed heartland of the township. The pasture not only allowed milk, meat, hides and wool to be produced, it also supported the oxen in the plough team. The meadow, along with holly boughs cut in the hollins, provided the fodder that allowed the core of the livestock to survive the winter, while woodland served the needs of communities whose homes, tools, vehicles, platters, temporary fences and heating were all largely based on timber. A lush and early hay crop generally depended upon proximity to water, so that hay meadows had their positions determined by the pattern of the watercourses. Pasture and woodland tended to be found in those places that were unsuited to arable farming – though as I have noted, the grazing of the stubble after harvest and pasturing in fallowing fields and furlongs greatly enhanced the contributions from enclosed pastures and commons.

In the centuries around the Norman Conquest, the woodland in Ripley will have existed on the higher ground of what much later became the Park, in the elevated northern section of the township around Birthwaite and on the flanks of the Cayton/Newton Beck where coniferous plantations now stand. Specific indications of places in medieval documents that are still meaningful places today may be hard to find, but there is one clear identification of the 'Wood of Ripley'. Around 1200, Nicholas de Cayton granted to the monks of Fountains his '. . . meadow in the Vale of Ripley and his land down from the said meadow between the wood of 'Rippeleia' and up from the wood as far as the road which goes from the said wood as far as the vill of Rippeleia, and then as far as the spinney north of the house of Ralph Blaber' (Chartul. of Fount., f. 98b). This shows that the Wood of Ripley overlooked the meadows of the valley of Ripley Beck and it probably surrounded part of the Whipley road, now seen as a holloway in the wood and parkland of the Park.

From the centuries following the middle of the Saxon period, woodland in England was in retreat, with the felling in Nidderdale reaching a climax in the thirteenth century. Clearances made in the pre-Conquest period are sometimes marked by *leah* and (sometimes) *_vait* names, as in Ripley, Whipley and Hampsthwaite. The grant of land by Nicholas de Cayton, just noted, also mentions his gift to the monks of his culture in Ripley which was called '*Wdhusum*' ('At the houses in the wood'). This place-name, which probably came into existence in the decades around the Conquest, is one that marks the start of the acceleration in woodland clearance that continued for a couple of centuries or so. In Yorkshire generally, the commonest of names associated with the post-Conquest clearances are those including –royd or –rode. In Ripley, common assarting name elements are Stubbings, denoting land covered with tree tumps, Riddings, and Ruddings, indicating land rid of trees and –breck, showing newly broken ground. Sometimes the name of an owner, perhaps the instigator of the assarting, is included in a place-name. Ripley had Edmundriding, Godwinriding and Elineriding. Other field names indicative of assarting were Buskie Close, Berestoch Stubbing (perhaps 'Swine pasture tree stumps') and Hellundebrec ('Cleared land by the little wood by the flat stone').

Although only a minority of assarts can be located with precision, it is clear that assarts and the associated farmsteads and hamlets formed a zone of colonisation that was sandwiched between the traditional ploughland on the one side and the wood or waste on the other. A good proportion of assarting names are clustered around the southern margins of the old common. This area must have borne the brunt of the clearance, with axes also ringing in Godwinscales around the boundary with Clint and on the higher ground. In later medieval times, the common occupied the northern part of the township above the assarting village of Birthwaite. A survey by W. Chippendale in 1752 mapped the large, squarish intakes which lay at the entrance to the old common and one can still recognise some of their features.

The quest to create new furlongs for the communal fields or privately tenanted closes placed pressures on the medieval resources of wood and common. It is highly probable that much of the waste existed in the form of wood pasture. This was a very widespread form of land-use, estimated to cover areas ranging from 24 per cent of the total area of Derbyshire, down to two per cent of that of upland Lincolnshire (Rackham, 1986, p. 21). It existed as pasture studded with useful timber trees that were pollarded above the reach of browsing deer or livestock. This allowed one piece of land to have a dual economy, with light timber, in the form of the poles that grew from the crown of the pollarded trunk or 'bolling', being produced among the grazings. The poles could be used in wattle fencing or walling, as fuel or as tool handles, while soft leafy growth, known in the locality as 'greenhews' or 'watter boughs' could be cut as fodder for cattle or deer. However, because of the delicate ecological balance between the wood and the pasture elements, the system was hard to maintain in the face of pressures from colonisation or grazing. Both the Forest and the monastic authorities attempted to conserve trees, while tolerating pollarding. Thus an inquiry into the customs of the Forest of Knaresborough in 1563 found that the felling of any tree older than 24 years would be severely punished '... yet nevertheless, it shall be lawful to and for the said customary tenants to take fire wood meet for fuel, growing upon the same, meet for repair, or to build his or their messauges [farmsteads] or ancient buildings there' (Greeves archive G19, Harrogate Library).

The area of commons with wood pasture in Nidderdale contracted severely during the Middle Ages. Following an extensive search in 1998 I could find only a tiny cluster of stunted oak pollards remaining, overlooking Leighton reservoir on the Nidderdale–Wensleydale interfluve (Muir, 2000). The following year my Ph.D. research student, Ian Dormor, found a group of ancient alder pollards quite nearby. These seem to be the sole survivors of a form of farming that covered thousands of acres in the dale. The last days of wood pasture in Ripley could be signalled by an agreement of 28 April, 1733 when John Ingilby of Ripley, son and heir of Sir John Ingilby of Ripley Castle agreed to sell 160 oak trees growing in the pasture of his tenant, Richard Pullan, at 'Banks' to one Ambrose Edwards of Co. Durham (Ingilby MS 2831).

Ripley, along with its neighbouring townships, Clint and Birstwith had a wonderful endowment of ancient oak pollards, many of them surviving into Victorian times and some still standing today. Most were hedgerow trees, but the oldest of all could be relics of wood pasture. The calculation of ages for ancient trees is difficult. They are invariably hollow, so that the oldest wood is not available for ring counting or Carbon-14 dating. The estimates of age have to allow for the species, its location, the high initial growth rate and the marked slowing in growth that takes place after the crown of a tree has reached its greatest extent. The oldest pollard found in Nidderdale according to a complicated system of age calculation (White, 1994) stands in Ripley Park and

FIGURE 21. The second-largest of the great oak pollards still standing in the deer park at Ripley. Its origins will lie well back in the medieval period.

FIGURE 22.
An old ash pollard amongst outgrown hedgerow thorns near the medieval marl pits.

was given an age of 891 – taking it into Norman times. It has a girth of over 29 feet (9 metres) (Muir, 2000).

Trees do not normally have documentary histories, but one Ripley tree, Godwin's Oak, does. Godwin, who gave his name to the western section of the township known as 'Godwinscales' (probably 'Godwin's summer pasture') was the grandfather of Avice, who lived around the start of the thirteenth century. To be regarded as a landmark, the oak must have been a prominent tree in Godwin's lifetime, so would probably have been at least 250 years old in 1150. Standing on the boundary with the Forest of Knaresborough, it was recorded in successive boundary surveys, being Godwy Ayc in 1234–50 (Ingilby MS 176) and Gawdewane oke in 1577. By 1767 it had fallen, though some of its remains were still visible. Almost certainly a pollard – and pollarding could have a rejuvenating effect – Godwin's Oak would have lived for a good 800 years (Muir, p. 106).

References to The Wood of Ripley show that there was some woodland as well as wooded pasture. Around 1200 it must still have been reasonably extensive, for Alan, son of Nicholas de Cayton was allowed to have 10 pigs in it each year (Ingilby MS 172). Thereafter, the references are to assarting rather than woodland or swine, though some woodland did remain. Hollybank Wood seems to be the Robert or Robter Wood of the medieval documents, and here there were hollins where holly branches were gathered. The less prickly branches from the tops of the trees were cut and stored in barns or helms (shelters), where they could be fed to starving livestock in winter to eke out the hay crop.

Routeways

The routeways of a locality are like a framework or a skeleton, and no understanding of the relationship between different settlements, or between particular farmsteads, hamlets or villages and their respective farmlands is possible without a reconstruction of a routeway plan for the period concerned. This is crucial – but it is also difficult. Different bits of a transport system are acquired at different times, while as some new routes are being gained, others are being shed. The transport network existed so that people could get about their commercial and social businesses. However, as villages were depopulated or as assarting or manorial policy shifted the centres of population and activity across the map, so the route pattern had to respond. Highways could be demoted to tracks and field paths might be combined to form a cross-country route. Reconstruction is not easy. Some clues may be provided by opportunities for relative dating, as when a road is seen to cut across an older field pattern. Air photographs, maps, documents and earthworks are all significant, while there is no substitute for walking a route on the ground. Only by walking might one find that there were paving stones beneath the turf, or medieval house platforms lining the roadside.

The oldest datable road in Ripley is the Roman road (though various lengths of lane or track could be prehistoric, for all that one can tell). It was almost certainly still intact and in use in early Norman times, and could have remained so into the fourteenth century. In 1365, the wagons, each drawn by ten oxen, that are recorded as having hauled lead from Greenhow, above Pateley Bridge, to Boroughbridge might have passed through Ripley (Jennings, p. 84). The 'high mountains and muddy roads' that plagued their journey were presumably higher up the dale. Parishioners from Godwinscales, Whipley and Clint would also have used parts of the Roman route when attending the first Ripley church, which survived until around the latter part of the fourteenth century.

Next in datable antiquity are monastic rights of way. The members of the great Cistercian houses, and particularly the lay brethren who served them, needed to be on the move. Their granges spanned areas very much larger than most farms, while their far-flung estates involved them in considerable travel. The monks were influential, well-connected, and at first greatly respected by their neighbours. They were, in short, far better placed than were most other people to negotiate convenient rights of way to link-up their scattered holdings. Between 1157 and 1173, William son of Ketel de Scotton, thought to have owned the smaller of Ripley's two manors, granted Fountains a cart road through his land from Hampsthwaite, through the middle of Owlecotes village, which lay in what became the Park, near Scarah Bridge, as far as the causeway on Dalbec (a crossing on Cayton Beck where the track forked, taking the monks to their grange or to the abbey). He received a mark in silver in return (Chartul. of Fount., f. 105). A little later in the century, Bernard the priest of Ripley, who owned the larger manor, gave the monks a road that was 40 feet (12.2metres) wide from Scarah Bridge to the same causeway (*Early Yorks. Charters*, I, p. 405). According to Jennings (1983, p. 82): '... the line of the road has been obliterated by the plough since the enclosure of the moor ... the causeway was probably on the road, now known as Red Gate Lane, which ran from the top of Cayton Gill, past Dole Bank, to Haddockstones Grange on the way to Fountains Abbey.' Around 1200, Richard de Ripley granted the monks of Fountains free transit through the vill for the movement of their cattle from the grange at Cayton to the grange at Brimham, a few miles up the dale (Chartul. of Fount.; Add MS 18276, f. 235d). Today, a deep, narrow branching holloway marks the descent from the old Ripley common to the ford leading to Cayton grange.

The road from Hampsthwaite bridge will have been the already ancient Roman routeway until the boundary of the vill was encountered, when a line heading NNE was adopted. As it ascends towards the watershed between the Nidd and the Ripley Beck it is now grassed-over marked by a holloway, while beyond the watershed and Monk Head Cross part of the course to Owlcotes is contained in a finger of woodland, where ridges in the woodland floor suggest that a strip or a group of ploughridges may have been donated to form the road. This occurred elsewhere in the vill, for a selion beside Newton Beck was granted to the monks by Robert son of Prince to provide the monks with access to their holdings in Newton (Chartul. of Fount., f. 105). In *c.* 1250–80, land called Shandekefalde ('Shand's cow fold') was mentioned as lying next to this 'way out to Newton' (Ingilby MS 180). The probable courses of the monastic rights of way are shown in Figure 23. In the northern part of the township, the track running northwards to Newton Hall; Green Lane, and Scarah Lane are apparently surviving fragments of these rights of way.

Other trackways can be traced that surely existed during the medieval period. In the Park a paved road was discovered during this survey beneath the turf blanketing a holloway. It ran roughly northwards from the Roman route at Sadler Carr (see below) in the general direction of the former village at Owlcotes, apparently on the alignment of a co-axial from the old field system. Various field tracks can be recognised amongst the ridge and furrow earthworks, and these will represent the access ways used by ploughmen and their teams

FIGURE 23.
A reconstruction of routeways in medieval Ripley. The locations of four wayside crosses, almost certainly dating from the period of Cistercian ownership, are shown.

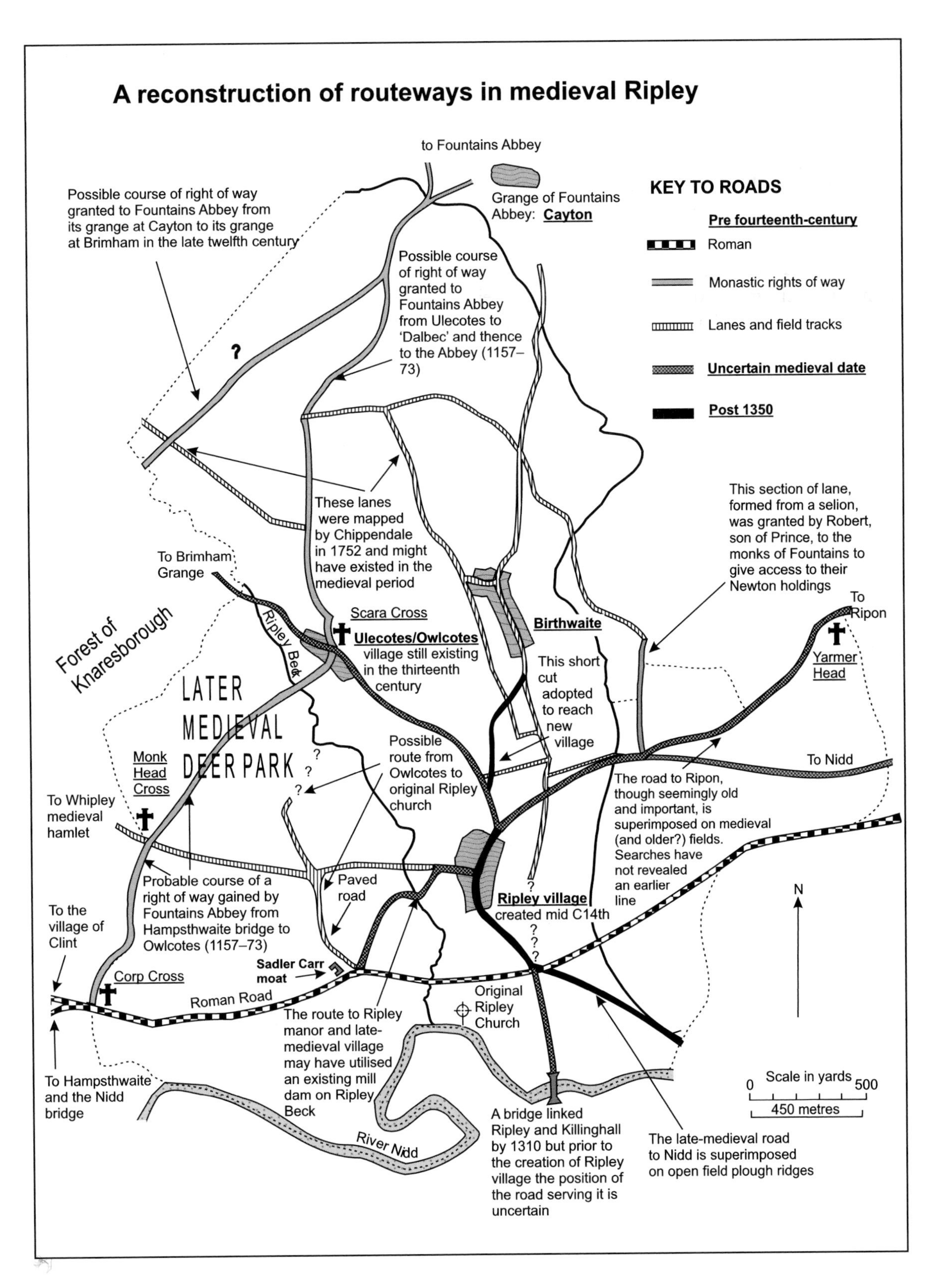
A reconstruction of routeways in medieval Ripley
to Fountains Abbey
Possible course of right of way granted to Fountains Abbey from its grange at Cayton to its grange at Brimham in the late twelfth century
Grange of Fountains Abbey: Cayton
KEY TO ROADS
Pre fourteenth-century
Roman
Monastic rights of way
Lanes and field tracks
Uncertain medieval date
Post 1350
Possible course of right of way granted to Fountains Abbey from Ulecotes to 'Dalbec' and thence to the Abbey (1157–73)
These lanes were mapped by Chippendale in 1752 and might have existed in the medieval period
This section of lane, formed from a selion, was granted by Robert, son of Prince, to the monks of Fountains to give access to their Newton holdings
To Brimham Grange
Forest of Knaresborough
Ripley Beck
Scara Cross
Ulecotes/Owlcotes
village still existing in the thirteenth century
Birthwaite
To Ripon
Yarmer Head
This short cut adopted to reach new village
LATER MEDIEVAL DEER PARK
Monk Head Cross
Possible route from Owlcotes to original Ripley church
To Nidd
The road to Ripon, though seemingly old and important, is superimposed on medieval (and older?) fields. Searches have not revealed an earlier line
To Whipley medieval hamlet
Probable course of a right of way gained by Fountains Abbey from Hampsthwaite bridge to Owlcotes (1157–73)
Paved road
Ripley village
created mid C14th
N
To the village of Clint
Corp Cross
Sadler Carr moat
Roman Road
Original Ripley Church
The route to Ripley manor and late-medieval village may have utilised an existing mill dam on Ripley Beck
To Hampsthwaite and the Nidd bridge
River Nidd
A bridge linked Ripley and Killinghall by 1310 but prior to the creation of Ripley village the position of the road serving it is uncertain
The late-medieval road to Nidd is superimposed on open field plough ridges
Scale in yards
0
500
450 metres

FIGURE 24.
The holloway of an old paved road in the deer park. The hunting tower lay just to the left and stone structures stood on the bank between the road and the tower.

of oxen as they moved their tackle between the scattered selions that made up their holdings.

Two problems remain. The first concerns the road that connected Ripon with places in Nidderdale – the medieval equivalent of the A61/B6165. Ripon was an early ecclesiastical centre, with Wilfrid becoming Abbot here about 660. Although the town's charter from Athelstan of 937 is probably a forgery, the place was influential in the pre-Conquest period and would have been a significant destination. However, the existing road, as marked by the A61-B6165 alignment, does not have extreme antiquity. An air photograph (NMR 17413/16) shows that it cuts the older co-axial field network at Yarmer Head, while it was also cut across the end of a medieval culture to the west of the crossing of the Newton Beck. As a result, ploughing in the severed section was realigned to run E–W across the original ridge and furrow corrugations. None of this would pose a problem were the traces of an older course for the route apparent, but no revealing holloways can be found. The road seems to have come into existence by 1200, for the village of Owlcotes and the documented beck crossing at Godwin's Bridge (Scarah Bridge) were both aligned upon it.

Secondly, there is the case of the bridge across the Nidd linking Ripley with Killinghall. The beautiful stone bridge that was built in a late-medieval style in the sixteenth century can be seen just downstream of the modern bridge. Various timber bridges, the first of which was recorded in 1310

FIGURE 25.
The deep holloway of the medieval road from Ripley to Whipley that was engulfed by the deer park. The oak pollards may be from the roadside hedges.

FIGURE 26.
The sixteenth-century bridge linking the townships of Ripley and Killinghall, both parts of Ripley parish. Some bridge or ford must have existed earlier in the Middle Ages but it is hard to discern the track that served it.

(Jennings, 1983, p. 84), preceded it. This would have been an important river crossing for the locality, the only one between Knaresborough and Hampsthwaite, but the question of how travellers from the northern side of the river approached the bridge without trampling across crops is a puzzling one. The bridge is served by a road from Ripley village, but this severs the end of a culture so that it cannot be particularly old, while the village may well not even have existed in 1310. If travellers reached the bridge by following the edges of the furlongs overlooking Ripley Beck then they did so without leaving any durable traces of a holloway. In terms of medieval earthworks, the approaches to the bridge are a 'blindspot' and have been levelled by later ploughing, while the current road from Nidd also cuts across medieval ploughland.

The inadequacy of the road system in medieval England was largely an expression of the inability of authority to create and maintain decent routeways. The matter was handled by the manor courts; throughout the length of the kingdom they threatened and implored a mass of peasants who plainly had other priorities than road building or mending. A late stage in the process is picked-up by the Ripley manor court rolls, as in 1628 when: 'A paine laid that Elizabeth Robinson of Skarey (Scarah) shall amend one pece of way at her wall side at Skarey end, being the way to the myll' (Ingilby MS 1608, no. 3, 18 April 1628). From the evidence of the rolls it seems that problems much more troublesome than the roads themselves were associated with the ditches that flanked roads and fields. Thus, in 1649, 'Thomas Burniston and Robert Mountaine shall scour their water-sewers going into the churchyard; Burniston shall also scour his at the marl pits' (Ingilby MS 16077, no. 18, 23 October 1649) and the next year: 'A paine laid that the occupiers of the comon peece, or Widdow Holdsworth, shall mak a ditch, and take the water out of Bursill {?} Laine, before Martinmas next, upon paine of every default iij*s.* iiij*d.* (3*s.* 4*d.*, about 16p) (Ingilby MS 1607, no. 1, 19 October 1650) ...

Reconstructing a medieval road network

This work is essential to any understanding of a medieval countryside, but the task is likely to prove very difficult. The work requires the dating of the different components in the existing network, the discovery of those roads, lanes and tracks that were once used but which have been abandoned and the ability to relate the reconstructed network to the other facets of the medieval landscape, like settlements, churches and markets. Remember that the medieval period was several centuries in duration. A place was likely to look quite different in, say, 1530, than in, say 1086. The medieval routeway plan was not fixed but fluid. It changed, with some routes being sloughed-off and different new ones created as people abandoned old destinations and discovered new ones. The following tasks should be attempted in recreations of the routeway patterns.

1. Consider all possible destinations, *whether these have survived or not.* For example, a subsequently deserted village would, in its day, have stood on a through-road and been served by minor lanes and tracks which linked it to fields and neighbouring settlements. Similarly, an abandoned church or chapel would once have been connected by routeways to its congregation.

2. Search the medieval documents for any direct or *indirect* references to roads and to journeys made to particular destinations. They may permit

an approximate dating of routes or give information about their condition, purpose or maintenance at different times, while mentions of the type of traffic could reveal the quality of the road and whether it could take wheeled traffic.

3. Study air photographs, not only to search for abandoned sections of roads, but also to establish relative dates, as when a road cuts an older field pattern or when an abandoned road is overrun by younger field ditches.

4. Where you have places that *must* have been served by roads – like former villages, bridges, fair sites and the old commons, which were accessible to all the people of the vill – try to discover the location of missing links that served them. If the roads cannot be discovered, try to reconstruct the field patterns. Roads would normally border furlongs. Where they cut across them, thus slicing-up the selions that were the units of tenancy, very special explanations must apply.

5. In the field, look-out for holloways and also probe any troughs, broad paths or green lanes with a ranging pole or metal pointed stick for traces of a buried paved or metalled surface. Double hedgerows can mark both boundaries and abandoned lanes. Roads and lanes were very attractive to medieval settlement. House or farmstead sites may be recognised as level roadside platforms, roughly of the dimensions of an old motor bus. House and croft/yard sites are suggested where the hedges or walls flanking a lane briefly sweep back from the track to allow more space for a road-sidehouse plot.

6. Also in the field, look for traces of contemporary field tracks among the ridge and furrow corrugations of the old open fields. They may be seen as slight holloways or wider-than-normal furrows between the ridges. Some will just have given access to selions within a furlong, but others were parts of longer routeways and some served local quarries or ponds. Their discovery helps to fill in the detail of the medieval countryside.

7. Throughout much of lowland England, waterways had a far greater commercial importance than land routes, with narrow barges reaching places far inland and shifting cargoes as bulky as building stone. The possibility of water links must always be considered and tracks could lead to vanished staithes or landing places.

8. Be aware of the potential of place-name evidence, like the stane ('stone') and street names associated with the Roman roads or the Saltergates or salt ways of the pre-Conquest centuries. The names can be colourful as well as informative, as with the Cut Throat Lane in Ripley's neighbouring parish of Bishop Thornton.

Settlement

One of the most striking facts to emerge from this study is the large quantity of deserted medieval settlements in the township, with villages, hamlets and farmsteads all being well represented. In England generally, the basic village pattern seems to have been set down in a time of tension and change during the Middle Saxon period, when the influence of the church was expanding, estates were breaking-up and open field farming was appearing. New villages were added to the pattern in the centuries that followed, continuing into the Norman period and even beyond. In Ripley, the situation appears to have remained fluid for much longer, so that the foundations of the modern pattern of settlement were not established until the years around 1400. During the preceding centuries, various potential focuses for village settlement were exploited, though none of them enjoyed lasting success.

The Late Saxon village whose traces were discovered beneath the medieval graveyard did not, for whatever reason, persist. Neither did the peculiar string of dwellings on the river terraces below. It is possible that the settlement associated with the original church drifted along the west-facing valley slope of the Ripley Beck in the direction of the present church. The slopes here are slightly corrugated with short plough ridges that run down to the beck, but between or cut into the ridges are faint traces of small house platforms. In two cases stone wall footings can be recognised, while in several examples probing with a ranging pole reveals rubble floors a few inches below the surface. The pattern of settlement here (see Plate 8) is too loose to be regarded as a village and the impression is of a group of small farmsteads, each in its own enclosure. The best example, lying just to the north of the place where the Roman road crosses the beck, has a house platform cut into the slope and an enclosure partly cut into the plough ridges. The pattern of loosely clustered farmsteads may have extended further north, but ambitious garden-making operations below the rectory by one of the incumbents will have obliterated all traces of it. Without excavation or pottery finds, any attempts at dating must be highly speculative, but the features seen could be associated with settlement that post-dated open field farming – but not by so very much. Perhaps they, or some of them, date from a century or so either side of the Norman Conquest?

The riddle of Hallstead

In the chapter that follows it will be shown that the village of Ripley as known today was the creation of the Ingilby family towards the end of the medieval period. But the question still remains as to whether some sort of settlement at or around this place preceded the creation of the planned village of the late fourteenth/early fifteenth century. The question seems to hinge on the interpretation of a place called 'Hallsteads'. The name means 'the Hall place' and may sometimes signify the sites of Late Saxon halls abandoned as the patterns of landholdings was transformed after the Conquest (S. Moorhouse, pers. comm.). This is easily the most commonly-mentioned name to appear in the early documents of the township and it stayed in use throughout the medieval period. Among the numerous mentions, about 1200 it is recorded as Hallestat (Ingilby MS 172), it appears as Hallestede in the mid thirteenth century (Chartul. of Fount., f. 88) and in 1411 and in 1414 as Hallstedes (Ingilby MSS 201 and 202). Such a name could very well signify the location of a medieval manor house, perhaps with attendant dwellings – but where was Hallsteads?

From the charters it appears to have been associated with the demesne – the land that the peasantry farmed directly for the lord – and with a block of land containing both ploughland and meadow. The lord of the larger manor had an arable culture or furlong in Hallstedes in the mid thirteenth century (Chartul. of Fount., f. 88), while about the same time meadowland in 'Hallsted' was the subject of a sale (Chartul. of Fount., f. 86). We should therefore be looking for a river- or beck-side location, where damp meadow lay close to well-drained ploughland.

The search for Hallstedes might easily have petered out for lack of further evidence, but for a sentence in *Parentalia*, a quirky geneaological history of the Ingilby family written by one Edward Clough in 1763–4 (Ingilby MS 3757). On page 20 he wrote: 'One favourite scheme which this gent [Sir Thomas Ingilby] pursued with much attention was the Improving of Ripley, by enclosing the common fields ... particularly ... Hallstedes (supposed Sorrowsykes) which was enclosed before 1413 ...' This information seems true in general but not in detail and reveals that Hallstedes had come to be known as Sorrowsykes ('Ditches of Sorrow' – had a drainage plan not worked?). Sorrowsykes can be traced. It is shown as lying in the general area of the present castle grounds and gardens on the Chippendale survey of 1752, while John Ingilby's planting diary for (?) November, 1782 notes: '... began to trench and (?) behind the garden wall in sorrowsykes' (Ingilby MS 2838). Hallsteads lay in the area of gardens around Ripley Castle. This strongly implies that the castle was built upon a manor house site that could already have been occupied for centuries, leading back into the pre-Conquest era. Whether any dwellings occupied by feudal tenants were also found here cannot be determined, although *most* of the site of the present village was covered in ridged ploughland until late in the medieval period.

Here, and in many other places too where village and township share the same name, the information supplied by documents is devalued by confusion concerning whether the clerks concerned were referring to the *village* or the *vill* or township. Usually, it was the vill that was the focus of interest, thus the mid thirteenth-century reference to '... a croft which lies beside the beck which runs through the vill of Rippelay, and abuts upon the meadow of the same monks called Bradeng to the south ...' (Chartul. of Fount., f. 88b) seems ambiguous. It might be saying that the beck runs through Ripley village, which would provide a date, a location and a confirmation of the existence of the village at this time. It is far more likely to be mentioning the beck that runs through Ripley township. Another reference, mentioned earlier, is more puzzling. Dating from around 1200 it notes the bounds of land held by Nicholas de Cayton and mentions a road running from the Wood of Ripley as far as the vill of Ripley (Chartul. of Fount., f. 98b). The road could be interpreted as the holloway running through the park in the direction of the castle. The 'vill' might be a village in the vicinity of the later castle; it might refer to the larger of the Ripley estates, or it might refer to a different village – Owlcotes – as being the village of the township of Ripley.

FIGURE 27. Owlcotes (*Ulecotes*) deserted village, Ripley. Following the discovery of the location of the lost village, information was passed to English Heritage. This resulted in air photography under perfect conditions, demonstrating the south-eastward extension of the settlement in an area of trees and rough grassland.

Owlcotes

Owlcotes, the 'Cottages Haunted by Owls', is a well-documented deserted medieval village although its site and dimensions were only discovered as a result of this survey. From the outset there is a clear documentary reference to the village. In 1155–95 Dolphin 'de Clotherham' granted land in the territory of Clotherholme to the monks of Fountains, and also '... an acre at the town-end between the road from Ripley to [Kirkby) Malzeard and Kexbeck ...' (Chartul. of Fount.; Add MS 18276, f. 235d). 'Kex' seems to be a diminutive for the personal name Kettil that was common in the Anglo-Danish north, so Kex Beck must be the Kettle Spring lying to the west of Scarah Bridge.

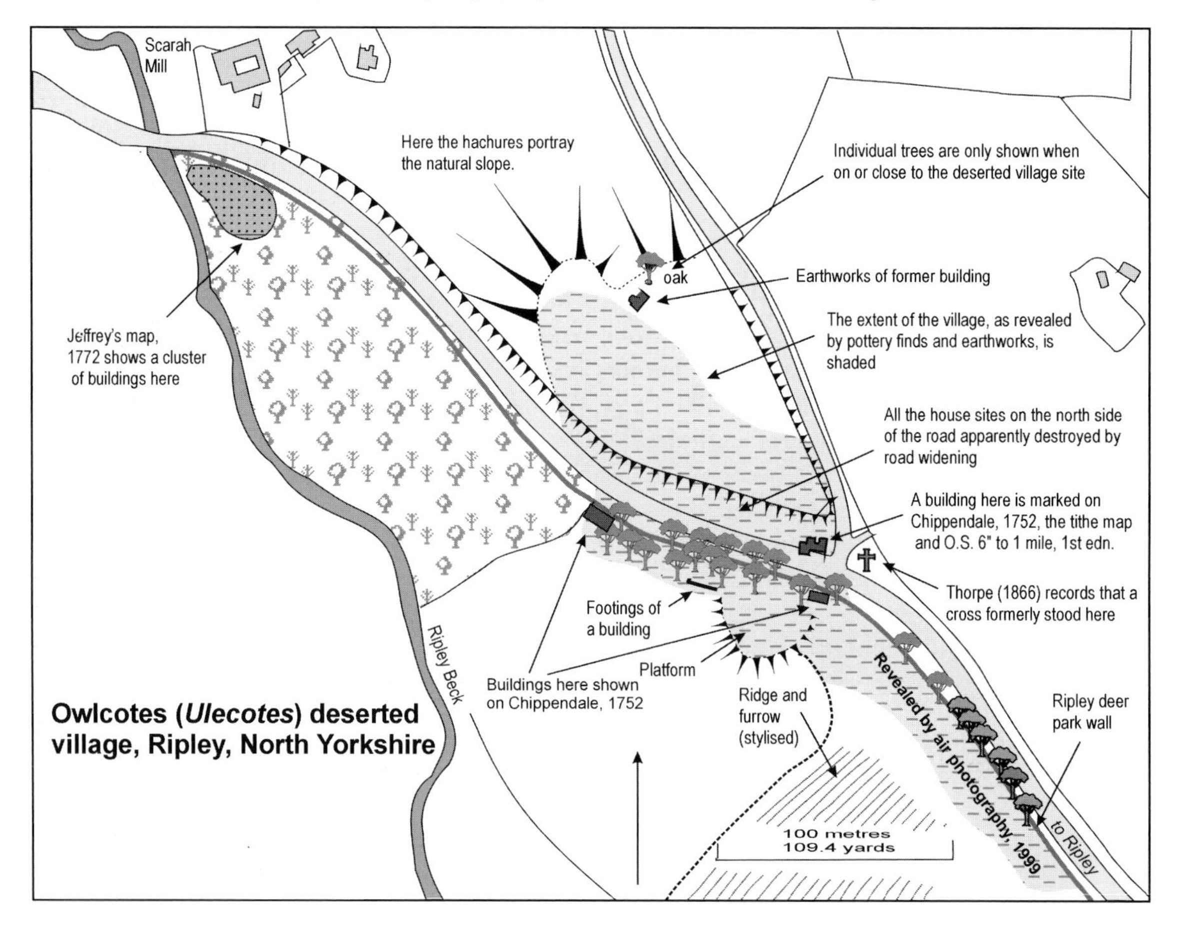

FIGURE 28. The deserted village of Owlcotes. The earthworks of closes and at least one dwelling can be seen on the upper side of the road just inside the deer park wall. Inside the park are the earthworks of ridge and furrow of widely separated ages, the narrow ridges probably resulting from steam ploughing in Victorian times.

This places the town end of Ripley in the western part of the township, where Owlcotes lay. (One might also presume that since this is the 'town' or village end, no other village existed in the township at this time.)

Abundant references to Owlcotes can be found for the period 1150–1300. The location of the village is established in the charter of 1185–95 (Chartul. of Fount., f. 257) which mentions the house of Thomas of Owlcotes (*Thome de Ulecotes*) as standing by the 'bridge of Ripley', i.e. Scarah Bridge. The earthworks show that the village continued eastwards along the road to Ripon for a distance of about 600 yards (549 metres). It was first recognised as a pronounced platform containing medieval pottery roughly opposite the junction with the Scarah Bank road, while the field in the angle of the junction also yielded fragments of pottery. After information concerning sites at Ripley was passed to English Heritage, several locations in Ripley Park were photographed from the air in November 1999 under highly favourable conditions. It was seen from the exceptional quality of the images that the village had extended further eastwards than originally imagined.

Owlcotes was a substantial place. The bulk of the pottery recovered dates from the twelfth and thirteenth centuries, apparently the heyday of the village, though a little later material shows the persistence of settlement here into the fifteenth century. The later settlement might only have consisted of a few freestanding dwellings. There is no clear-cut reason for the desertion of the place – though it did lie within the territory of the late-medieval park. The arrival of the Black Death in 1349 might account for the desertion – though most villages that were depopulated by the Pestilence were repopulated within a few decades. It seems more likely that the lord preferred to have his tenants somewhere else. Desertion was not necessarily sudden or total. A 'toft and croft', which implies a dwelling and a plot of land, at 'Houlcotes' were recorded in 1314 (Ingilby MS 190); lands in 'Ulcotes' were mentioned in 1434 (Ingilby MS 219). However, although the Chippendale survey of the mid eighteenth century shows two buildings standing, perhaps coincidentally, on the deserted village site, there seems no doubt that by the end of the Middle Ages Owlcotes was well and truly dead.

Sadler Carr

The name of this place (perhaps denoting a swampy piece of woodland owned or occupied by a saddler/someone called Sadler or perhaps more convincingly given the status of saddlers, referring to a sallow carr or willow swamp), does not appear in the old records and must be relatively recent. The settlement here, a moated homestead, must be old for the date range for these features '... extends from the late twelfth century into the post-medieval period, with a peak period for construction between the late thirteenth and mid fourteenth centuries' (le Patourel, 1981, p. 7). The moat took the commonly adopted plan of three sides of a rectangle. To the extent that the earthworks were defensive, one moat guarded the approach from the adjacent Roman road, one faced the paved causeway leading from the road towards Owlcotes, and the shorter

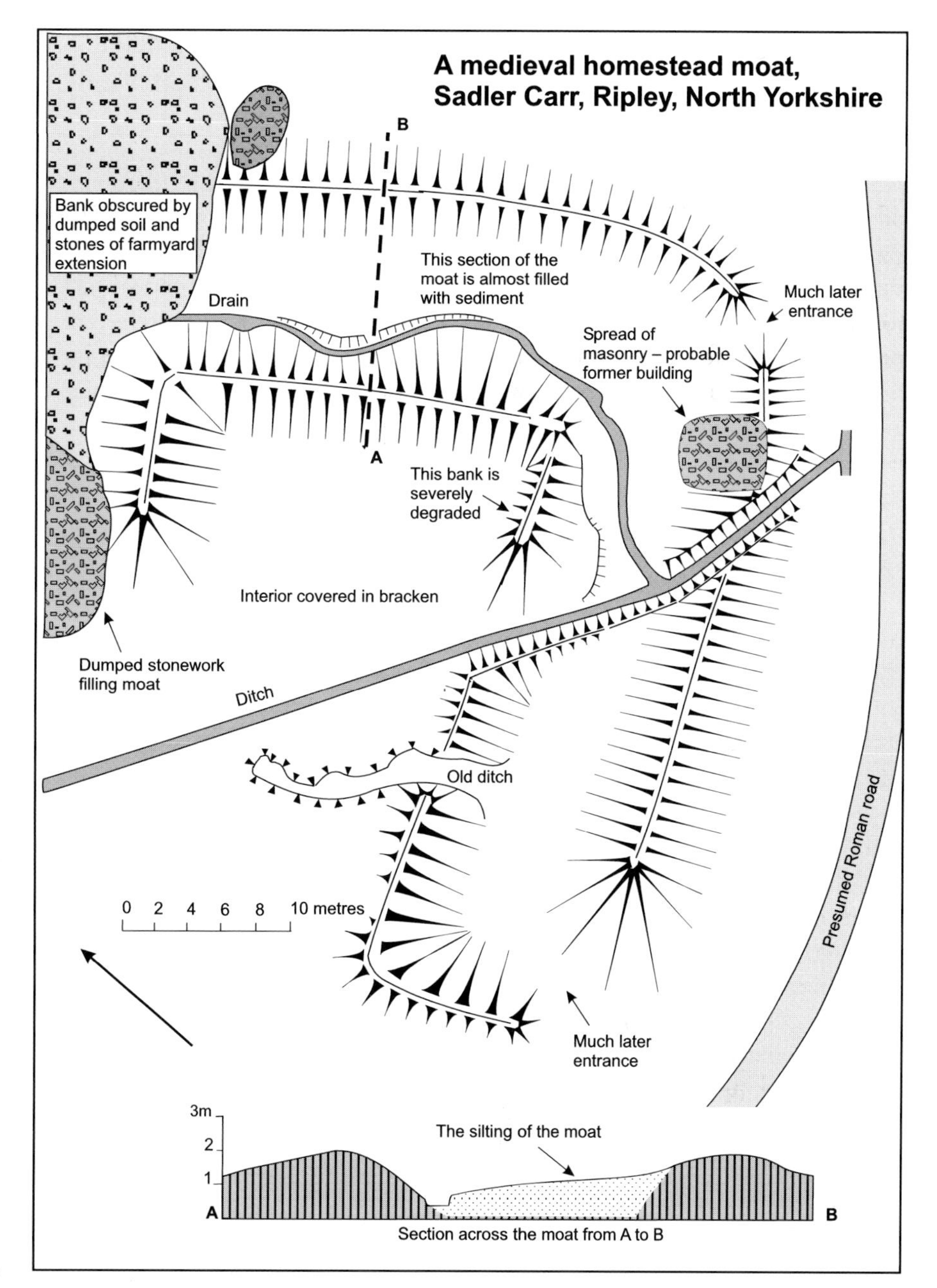

FIGURE 29.
A medieval homestead moat, Sadler Carr, Ripley. The site is overgrown and cut by later ditches. Extremely active silting of the north-eastern section of the moat by a small drain has resulted in the almost complete filling of the moat, although the inner and outer dump banks still stand proud.

third moat faced the north-west. Recent waste dumping has had an effect on the monument, while silt deposited by a stream entering from the north-west has largely filled the north-east facing moat, which is only recognisable from its upstanding inner and outer dump banks formed of material excavated from the moat.

Originally, the moated platform will have contained a dwelling and probably a few outbuildings. The top stratum of occupants of places such as this were gentry who could not quite aspire to castle-owning status, while the lowest stratum comprised reasonably prosperous freemen, farmers and yeomen. It is very likely that the owner and occupant of the site is recorded as a grantor or witness in one of Ripley's numerous medieval charters – but as we do not know the medieval name of the site we must fall back on speculation. If we assume, as I have argued, that the de Ripley manor house, the headquarters of the larger of the two manors, was at the present castle site, then Sadler Carr would not have been the main family residence. It would be well-placed to serve as the headquarters of the smaller manor, comprising Godwinscales, but most of this territory fell under monastic control by the mid thirteenth century. It is more likely than not that the moated site was excavated and first occupied by a minor member of the local gentry between about 1275 and 1350. Various candidates could be put forward – a minor member of the Ross family? A younger de Ripley brother? A de Scotton? A descendent of Godwin, who himself probably lived too early to be the creator? A relative of Huckeman, whose family controlled the neighbouring hamlet of Whipley? No clear claimant can be identified.

The occupation of the Sadler Carr site continued well after the close of the medieval period. In his celebrated atlas of itinery maps of 1674, Ogilby annotated the route from York to Lancaster as it passed to and through Ripley in the following terms: 'Nid Ch. And Hall on the right, 4f., and some lime pits on both sides, at 22–6 to Ripley, of 3. F., seated likewise on the Nyd, and having a market on Fridays, Dark Hall right; by Clynet, a village to the right ...' The positioning of 'Dark Hall' makes the label seem to allude to Sadler Carr rather than to Ripley Castle, though the castle gatehouse was an imposing roadside landmark. Whatever this seventeenth-century Dark Hall may have been, the moated platform carried a building until the nineteenth century. In 1866, Thorpe described an old barn and outbuildings and suggested that the ruin had 'at some remote period been a residence of some importance' (1866, p. 106). There were massive oak timbers with mortices and fittings suggesting that they had been part of a mansion, and the remains of pavements and foundations among the surrounding defences, though the moat itself was faint yet still unmistakable. Were these the remains of the original homestead, probably built in the thirteenth or fourteenth century, or was it a later high status building – and if so, whose? The fine old oaks that Thorpe found in the vicinity have since disappeared. A little later, in 1871, Grainge wrote that: 'Dark Hall does not exist at present, but has been removed within living memory' (p. 34). Today, a scatter of masonry on the outer dump bank near the eastern angle of the moat is all that remains to suggest a former building. Nearby is a large glacial erratic boulder, its top apparently hollowed as a basin, a place for leaving alms for lepers, or some other function.

Birthwaite

The process of village foundation in England continued for centuries after the Norman Conquest, with the later additions to the village flock often being small places that were established near the frontiers of cultivation in the more marginal environments. Birthwaite seems to have been a typical assarting settlement. The name is first encountered in the thirteenth century and the settlement grew beside a lane that led to the common of the township at the place where the lane met the common and where assarting and intaking were active. Birthwaite presents challenges of definition: was it a hamlet or a village and should what remains be regarded as a deserted or a shrunken village? At its peak, the settlement would have had between six and twelve dwellings, perhaps more, and its size and its well-structured lay-out place it in the village category. Only two farmsteads or old farm houses remain, and the village seems more lost than shrunken. The lay-out was of the 'Y'-shaped form that is so characteristic of villages in the region. On reaching the margins of the common, the lane forked, accommodating a small, triangular green in the angle of the arms. Dwellings will have lined the stem of the 'Y' and the outer sides of its limbs, facing towards the green. By the mid eighteenth century the shrinkage was almost as complete as today, though as well as the two farmsteads enduring today on the eastern side of the lane, there was another on the western side. The decline of the little village might have begun before the medieval period was over, even though pronounced house platforms can still be recognised on the western side of the lane.

Evidence that Birthwaite was a small village with a distinct structure and a population in the early seventeenth century comes from an entry in the Ripley

FIGURE 30. The site of the deserted small village of Birthwaite looking towards the former common. The green, still identified as a green in the seventeenth century is accommodated in the fork of the road and dwellings stood by the roadside.

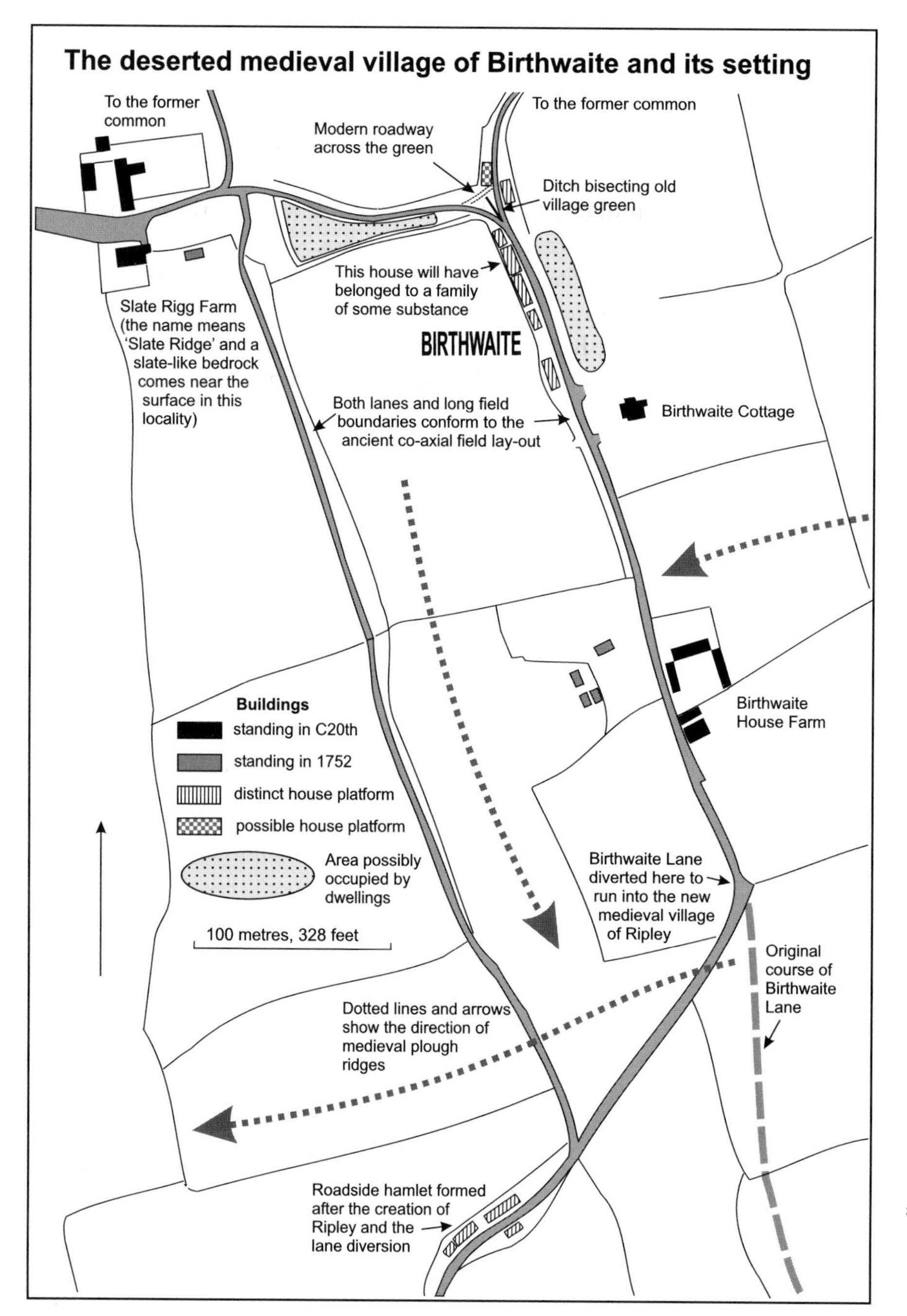

FIGURE 31.
The deserted assarting village of Birthwaite. The broad nature of the zone between the field ditch and the track directly to the west of the little triangular green suggests that it may have been incorporated in the village lay-out, either as house plots or as an additional green. Any earthworks here are very slight.

manor court rolls: A 'paine' laid '... that the inhabitantes of Birthwait shall appoint which is the right course of the water which cometh of Birthwaite Greene, and cause the same to rune downe the same' (Ingilby MS 1607, no. 14, 13 May, 12 Js. I). (Commonly in such documents the accused are said to be 'amerced' or 'in the mercy of the court'. Here the term 'paine' is consistently used. A total of 55 rolls survive, and although they do not begin until the late sixteenth century and continue to the late seventeenth century the style remains medieval and, with the exception of the 'pains' and some terms which were difficult to translate into Latin, the clerks still employed Latin, the language of the law.) The reference to a green identifies the triangular island in the tracks at the head of Birthwaite, while the instruction that the inhabitants should agree upon a proper course for the stream flowing by the green implies a meeting between a number of heads of families.

The Pallisers, described as yeomen in 1668 (Ingilby MS 1034) were the leading family of Birthwaite. Their farm, clearly on assarted land with field names like Wood Close, Buskie Close and Intake Close, consisted in 1675 of ten acres (four hectares) each of ploughland and meadow and twenty acres (nine hectares) of pasture (Ingilby MS 1044). They were said to hold their lands from the Ingilby family in Ripley by payment of a red rose on Midsummer-day and by carrying the boar's head to the high table during the twelve days of Christmas, in commemoration of an Ingilby ancestor's reputed feat of saving the King from an enraged boar whilst hunting in Haverah deer park. The hall of the Pallisers at Birthwaite was apparently demolished in the mid nineteenth century. It was described as: '... one of those antique yeoman dwellings with open thatched roof, and spacious unenclosed chimney ...' (Thorpe, p. 114). In the garden wall grew the scarlet rose that yielded the rent.

Hamlets and farmsteads

Ripley lies at the limit of England's lowland zone, where villages generally (but not always) predominate, while the upland zone, a land of dispersed hamlets and farmsteads, begins just a couple of miles further up the dale. Today, dispersed farmsteads are not numerous and the village of Ripley dominates the township. In medieval times, particularly before the mid fourteenth century, there was a great deal of dispersed settlement and Owlcotes was balanced by numerous hamlets and out-lying farms. Some of remains of this dispersed settlement must have been destroyed by ploughing or obliterated by road-widening or the concreting of farm yards, though this survey has still discovered a great deal. The most typical form for a medieval hamlet on the basis of our evidence was as a loose group of farmsteads strung-out along a road. Such a hamlet existed on both sides of Hollybank Lane, just to the west of Lodge Hill Wood. On the north side of the lane, the roadside platforms that carried narrow, elongated buildings can be recognised, with one more substantial building being set back in a close. Just opposite, the wood bank seems to have been looped back from the roadside to allow for the presence of a pre-existing dwelling, suggesting an early date for the settlement.

A hamlet of four buildings at the dog-leg in the lane leading up to Birthwaite probably existed from a later stage in the medieval period through until around 1800. The house platforms on the northern side are particularly well-developed and preserved, while a tree growing on one platform after its dwelling was abandoned has recently fallen, bringing up blue and white pottery, probably Dutch from the years around the start of the nineteenth century, embedded in the earth around its roots. The northern platforms are littered with eighteenth-century black, glazed pottery of the type known as 'Cistercian' ware. This might suggest that the abandoned platforms had been used for dumping refuse around 200 years ago – but this would not explain the presence there of older 'Cistercian' ware of the sixteenth century. A third hamlet of a more puzzling kind lay in what became the deer park, beside the paved track leading from Sadler Carr to Owlcotes, with some small buildings apparently standing on a bank and being flanked by holloways on both their eastern and western sides.

Isolated farmsteads and the earthworks of what appear to be a farmstead with one or more associated out-buildings can be recognised in various places, and some certainly stood isolated amongst the furlongs of the open ploughlands. Whatever the situation in the heartlands of open field farming in the Midlands, at Ripley several farming families must have lived outside either of the villages. At least four different sites can be recognised as earthworks in the medieval ploughland in the angle between Birthwaite Lane and the road to Ripon. In the east of the township, where parish boundary and Roman road coincide, the building platform and earthworks of a farmstead have overrun the northern ditch of the road. Without excavation a problem with most of these earthworks concerns the difficulty of differentiating between dwellings and other farm buildings. Barns and also shelters known as 'helms' were part of the landscape of medieval farming, and a few documentary references to farm buildings occur, as with 'Carterstabill cow house', 'laylathe' (laithe=barn) and 'stone Byrnylarr' mentioned in 1416 (Ingilby MS 208). In the Park directly facing the castle across a lake are the earthworks of a farmstead with two platforms cut into the slope, one presumably for the house and a smaller one next to it for an out-building. On the bank that separates the little complex from the holloway of the old road to Whipley are a row of slight mounds which were stack stands. A date before the creation of the deer park and probably before the village seems liklely.

Industry

Ripley lay in an area associated with a water-powered textile industry and the manufacture of armaments and the availability of watercourses assured an industrial element to the local economy. Probably the earliest involvement with industry that can be detected comes from a place-name, 'Walk Mill Ing', that was plotted on a map by James Powell of 1838 and other walk mills were recorded at Ripley in the fourteenth and fifteenth centuries (Ingilby MS 211). The same land, 'Walkemill Inge' was mentioned in the manor court roll in 1631 (Ingilby MS 1608, no. 8). The walk mill meadow that Powell mapped flanked the Ripley Beck in a section of the stream just south of the castle that had a long association with industry. Walk mills seem generally to have dated from an early phase in the medieval period, before the adoption of water-powered fulling mills, when the cloth was trampled in the water by human feet rather than being beaten mechanically. Numerous later mentions of milling here can be found, while it is probable that when a medieval link road from the Roman route at Sadler Carr to the main manor site was developed, the top of the mill dam was exploited to allow a dry-footed crossing of the Ripley Beck.

A document of 1421 mentions a fulling mill of Ripley that was leased for 23*s.* (about £1.15), but this was certainly located towards Knaresborough, at Scotton Banks (Ingilby MS 211). However another of the same year (Ingilby MS 210) mentions a lease from the rector of Ripley of a water mill with an adjoining close called 'Mill Close' that contained a house. This will have been to the west of the church and directly south-east of the dam/bridge on the beck, where a mill building was plotted by Chippendale in the mid-eighteenth century. Mill Close was mentioned again in 1523 (Ingilby MS 1064) when it contained timber and was said to lie between the parson's orchard and the park. Adjoining it and lying between the mill beck and the park was another piece of ground called 'Barkhousegarth'. The reference to a bark house reveals the existence there of a tanning industry which employed oak bark in the tanning process. The industrial use of the site persisted until well into the eighteenth century, while elsewhere on the estate, on the Nidd at Killinghall, John Oliver leased a piece of riverside ground on which to build a watermill and maintain it for 21 years from Sir John Ingilby in 1792 (Ingilby MS 2135). Beside the site of the old mill building is a faint circular earthwork about 26 feet (7.9 m) in diameter, perhaps the last trace of a donkey-powered engine used in some milling function.

In the sixteenth century there was a corn mill just across the River Nidd at Killinghall, while Ripley's other neighbour, Nidd, acquired a fulling mill and dam at the start of the century which caused water to back up to the Killinghall mill. A corn mill was established at Scarah, where the Thornton or Ripley Beck entered the township, by this century.

The digging of marl, used to sweeten farmland, at Ripley had more than just a local significance and has left massive earthworks. A narrow band of Magnesian Limestone was exploited close to the place where the road to Nidd crosses the Newton Beck. It was certainly dug in medieval times, and when Ripley's two manors were united in 1494 (when Sir William Ingilby exchanged his land at Knapton, Malton for the estates of Fountains Abbey at Ripley) the agreement allowed Fountains to take marl from the South Field for use on its granges of Morker and Haddockstones (Fountains Abbey Lease Book, entries 225 and 226). In 1538 the pits were being actively worked and Robert Rypley, a yeoman of Birthwaite, was given the right to take away marl (Ingilby MS 1066). Ogilby recorded the marl pits in 1674, while Thorpe believed that the digging of marl continued into the early nineteenth century, when a Mr Thomas Crowther, a local man, was the last occupier. He also implied that during the later stages of their exploitation, kilns were employed to convert the stone into lime. The earthwork of a dam, probably of medieval date, on the beck directly below the pits has been detected, though it is not known whether this served a mill or was used to flood the meadows upstream and stimulate the grass into growth. The best of the Magnesian Limestone served as a top class building stone, displayed at Fountains Abbey and other prestigious buildings. At Ripley it was exploited as a fertiliser and the churches and castle employed a purplish gritstone obtained in the Park. Other strata available locally yielded bakestones (for baking on) and stones for scouring work could be obtained from the trackside quarry in Hollybank Wood.

Finally, what of the people of Ripley? In popular mythology medieval peasants tend to be regarded as cretinous dwarfs. The vandalisation of the original churchyard in the laying of a water pipeline in 1991 gave rise to a rescue excavation and an examination of the skeletons of Ripley people, with 124 skeletons being exhumed (K. Cale, personal communication). A sample of these were taken to Bradford University for analysis and then re-buried in the current churchyard. About ¼ of the burials were those of infants, reflecting the high rate of infant mortality at this time, with children often being interned beside their kinsfolk. One mother presumably died in childbirth and her baby was buried with her. The soils in the graveyard above the Sinking Chapel were conducive neither to the preservation of wood nor bone, while the storing of manure on the land above accelerated the decay, but in a few cases the survival of nails suggested that coffins had been used. Bodies were aligned east-west, facing to the east in the conventional manner. The Ripley people

FIGURE 32. The cratered topography of the medieval marl pits, where limestone was dug for spreading on fields to sweeten the soil.

FIGURE 33.
The earthworks of an ancient dam and pond on the Newton Beck below the marl pits. No record of their function can be found, though the local field names included a reference to a 'stank' or pond.

of the twelfth and thirteenth centuries were not short, the growth in height being a very recent phenomenon related to our greater mobility which has envigorated the gene pool, and improved nutrition. Apart from the dangers associated with childbirth and the plagues and famines which periodically ravaged medieval communities, common disabilities concerned tooth decay, which could prove fatal and was intensified by the abrasive effects of grit entering the bread from the milling process, and diseases associated with exhausting work in bad weather, like arthritis. The average span of life in medieval Ripley will have been shorter than today, with the average expectancy of life dragged down by the high infant mortality. Whether the *quality* of life was worse than today is much less certain.

No manor court role is available for the medieval centuries, but the record of 'pains' in the seventeenth century shows the continuation of characteristically medieval offences. The people do seem to have been prone to fighting, if not particularly disorderly, and sometimes stubborn and negligent in their performance of their responsibilities to the community in areas like ditching. Typical crimes include not controlling their livestock; selling sub-standard ale; breaking hedges; stealing materials from the common; not clearing or regulating their ditches and sewers; neglecting their landlord's buildings, and taking part in fights of varying natures and sizes.

References

Austin, D., 'Medieval settlement in the North-East of England – retrospect, summary and prospect', in Vyner, D. E. (ed.), *Medieval Rural Settlement in North-East England*, Architectural and Archaeological Society of Durham and Northumberland, Durham, 1990, pp. 141–50.

Bishop, T. A. M., 'The Norman Settlement of Yorkshire' in Carus-Wilson, E. M. (ed.), *Essays in Economic History*, vol. II, 1962.

Chibnall, M., *The Ecclesistical History of Orderic Vitalis*, Oxford University Press, Oxford, 1969.

Clough, E., *Parentalia*, Ingilby MS 3757, 1763–64.

Darby, H. C. and Maxwell, I. S., *The Domesday Geography of Northern England*, Cambridge University Press, Cambridge, 1962.

Grainge, W., *Harrogate and the Forest of Knaresborough*, John Russell Smith, London, 1871.

Jennings, B. (ed.), *A History of Nidderdale*, 2nd edn, Advertiser Press, Huddersfield, 1983.

Muir, R., 'Pollards in Nidderdale: a Landscape History' *Rural History*, 11 (2000), pp. 95–111.

Palliser, D. M,. 'Domesday Book and the Harrying of the North', *Northern History*, 29 (1993) pp. 1–23.

Le Patourel, H. E. J., 'Moated sites in their European context' in Aberg, F. A. and Brown, A. E. (eds), *Medieval Moated Sites in North-West Europe*, BAR International Series 121, Oxford, 1981.

Rackham, O., *A History of the Countryside*, Dent, London, 1986.

Thorpe, J., *Ripley: Its History and Antiquities*, Whittaker, London, 1866.

CHAPTER FOUR

A Time for Changing

For people living in England, and also in the civilised world far beyond, the fourteenth century was a time of great traumas. They stood by, largely helpless, and saw the age of growth and expansion that had lasted for longer than the span of communal memories decay. For the folk of northern England the first hints of a worsening in the climate were soon accompanied by the terrors of Scottish invasion. At the mid-point in the century, the Pestilence or Black Death arrived, exterminating whole families, undermining communities, destabilising the feudal order and leaving a legacy of antagonism between the buyers and sellers of toil in a vastly depleted labour market. All these things must have been experienced in Ripley, but, quite remarkably, the fourteenth century, with all its horrors and failures, was the very making of Ripley. It could all, so easily, have been different. The accidents of chance encounters, a breakdown in marriage negotiations, odds of around evens that one partner or the other would fall victim to the Pestilence – any of these eventualities, and others too, could have wrecked the union that brought transformation to the township.

The background to change

The Scottish raids appear to have been experienced directly in Ripley. The Scottish victory at Bannockburn in 1314 and the weakness of the English monarchy left the north exposed to invasion by the war bands of Sir James Douglas. In 1318 the raiders torched Knaresborough, burning down seven-eighths of the dwellings. The surrounding area was ravaged, the cattle rustled and the villages razed. The damaged townships appealed for reductions in tax, claiming that they had been impoverished by the raids. The Abbot of Fountains sought compensation on behalf of his unfortunate tenants, and the gift of a market charter free of charges for Pateley Bridge seems to have been the reward. We cannot know where or how Ripley was struck, but Owlcotes, sitting astride a significant routeway, must have been too obvious a target to be missed. The tax assessment of the church at Ripley, mainly funded from tithes paid by parishioners, was reduced from £23 6*s.* 8*d.* (£23.33) to £10, a sure sign that the locality had been injured by the Scots (Jennings, 1983, p. 53). The Scots may have returned the following year, at a time when the onset of wetter, more cyclonic conditions was causing diseases to erupt amongst the herds and flocks. These disquieting events did not put an end to the expansion of farming. Axes still rang in the remaining stands of woodland, but far harsher times were in store.

The Black Death appeared in Nidderdale in 1349, by which time it had spread along the Mediterranean seaways to devastate the countries of continental Europe and had erupted in southern England. It was carried by fleas infesting the black rat population, with the infected fleas often adopting human hosts when their rodent hosts expired. It was almost invariably fatal and caused terror in both urban and rural populations that were already well familiar with famine and the terrible bacterial illnesses that swept away the people weakened by hunger. The Pestilence arrived in a northern world where wells and streams

FIGURE 34. The fragility of settlement in the Ripley region. The region is an archaeological showcase of lost, shrunken and shifted settlements.

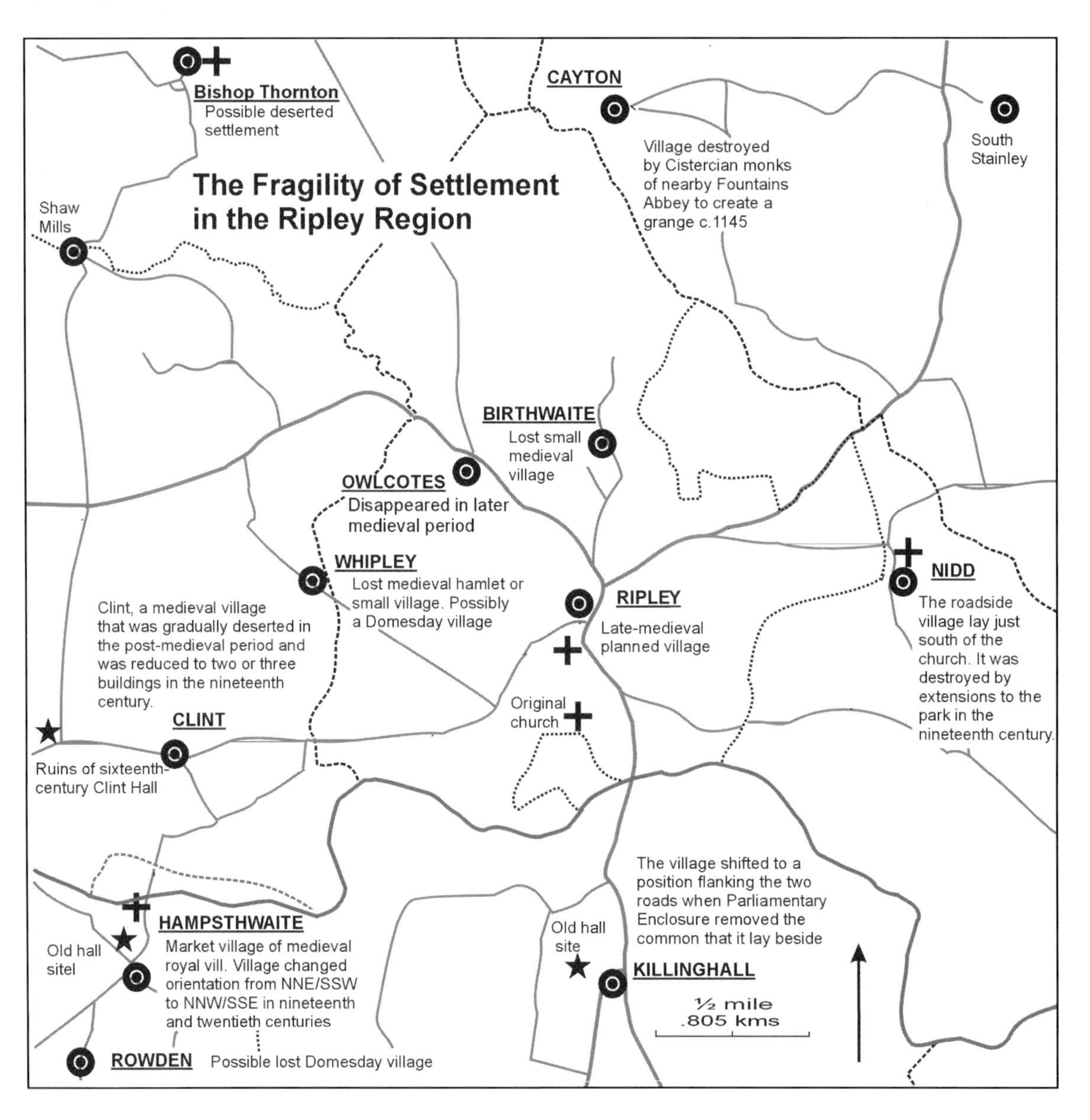

fouled by cess pits and middens were commonplace. It exterminated communities in the stinking alleys and villages and in the sanitary monasteries with fresh water systems alike. Nobody knew the cause of the illness or how to avoid contamination.

The great majority of the people who were killed by the Pestilence were ordinary feudal tenants whose deaths were considered inconsequential and unworthy of being recorded. Consequently, the levels of the mortality must be inferred from the meagre documentary sources. Jennings thought that in the adjacent Forest of Knaresborough, between 45 per cent and 50 per cent of land holdings were vacated through death in 1349–50 (p. 89). A small minority of vacancies would have been through death by causes other than the Black Death, but still a mortality rate of over 40 per cent would have resulted from the Pestilence itself. This accords with common estimates of a national death rate of between one third to a half of the population from the initial onslaught of the disease. The advancing frontier of assarting came to a halt. Where they had survived, the widows and children took over the empty holdings in the better farmlands, but the new lands on the upland margins of agriculture will eventually have been overrun by bracken, hawthorn and briars.

The old lords of the larger Ripley manor, the de Ripley family, took their name from the township, implying fairly humble roots. Their origins are unknown, but individual family members are encountered in a succession of charters. They were established at Ripley within a few decades of the Conquest as subordinates of the de Ross lords and may have been minor continental supporters of the Conqueror. William, whose name suggests such affinities, is recorded early in the twelfth century (Thorpe, 1866, p. 27) and we know of Bernard, who combined the roles of lord of the manor and priest and his brother, Richard, who lived at the end of the twelfth century, and Thomas, son of Roger de Ripley, who bequeathed his holdings in Godwinscales to the monks of Fountains around 1234. Under its namesakes, Ripley must have been a conservative, even a rather backwoodsy sort of place. As the middle decades of the fourteenth century approached, the last of the indigenous aristocracy was the heiress, Edeline Thweng. Perhaps the most momentous event in the history of the locality was the marriage, some time around 1330, between Edeline and Thomas Ingilby (1290–*c.* 1369). This marriage gave the manor to the family that would, in the course of a few generations, redraw the township map and which would direct the course of development for centuries and still be in residence today.

The legacy of the medieval Ingilby lords

The year and place of the marriage of Thomas and Edeline are not known and the early history of the Ingilby family at Ripley is confused. One may read published accounts by Thorpe (1866) or Lancaster (1918), download a genaeology from the Internet or consult with local experts and in each case a different version of the family tree is provided. Since this is a study of

FIGURE 35.
The tomb effigies of Sir Thomas and Edeline. Their marriage set Ripley on a new course of development.

landscape rather than family history, the genealogical issues are not explored in detail. The most significant fact must be that the Ingilby family became established in what was, if not a backwater then at least a rather peripheral part of a distressed and unfashionable part of the English kingdom. They introduced a worldliness, an awareness of the features associated with the commercially active estates and a more sophisticated sense of lordship to the township. The Ingilby family arrived at Ripley late in the medieval period, but before it was over all the most notable features in the local landscape had been transformed. Their geographical background appears to have been in Lincolnshire and then at Haughton-le-Skerne near Darlington, while Thomas built a successful career in law, becoming an advocate in 1347, a member of Parliament for Yorkshire in 1348, a Judge of Assize in 1351 and a Justice of the King's Bench in 1361. Towards the end of his life he was knighted. The higher stages in his career were bracketed by the arrival and the almost equally horrendous return of the Pestilence. During this time he worked in a kingdom riven by the social tensions which resulted once the peasant masses realised that the massive depletion of the population had made their labour scarce and valuable. Given this background of ravaging, disease and unrest, one might have imagined that the new lords at Ripley would have had their views coloured by pessimism and have acted with caution and circumspection. This was far from being the case: they did not entrench, they transformed.

The market and roads

Before his death around 1369 and his internment in the old church, Thomas may have initiated the great changes. In 1356 he obtained a grant of free warren

FIGURE 36.
The cross and stocks at Ripley. The gaining of a market charter was one of the first acts by the village founders to put Ripley 'on the map'.

for his demesne lands in Ripley, Flask, Amotherby and Hutton Wandesley. This right to hunt small game (granted in a sensitive locality bordering a royal Forest) may not have had great repercussions for the Ripley landscape, though the achievement of the following year must have done. His record of promotions suggests that Thomas enjoyed the favour of Edward III, so that the grant of a Monday market and annual fair lasting for three days on the eve of the day and morrow of the Assumption of the Blessed Virgin at Ripley may have been made without the considerable financial outlay that such privileges normally required. The question of where this market was held is an intriguing one. I had imagined (Muir, 1998, pp. 11–12) that the acquisition of the market must have been the occasion for shifting the village from the vicinity of the original church to its present location. However, on closer scrutiny there is no convincing proof that in 1357 a village existed at either location – although within a few decades the market square would be the centrepiece of a new, planned village. It might have been held in or beside the old churchyard – though in terms of attracting passing trade this would have been something of a *cul de sac*. It could have been held near to the manor house, somewhere close to the later castle. No other potential locations spring to mind.

Lords were keen to secure the rights to hold markets and fairs because they encouraged commercial vitality on the estate and produced trickles of revenue from the fines imposed for trading irregularities, while kings were always pleased to profit from the sale of the privileges. Consequently, many regions were over-stocked with competing markets, of which only a minority ever amounted to very much. Since at least the start of the twelfth century Ripon had a Wednesday market and two important fairs; from the start of the thirteenth century Knaresborough had a market, which also came to be held on a Wednesday, and the borough gained a fair in 1304; Hampsthwaite gained a Friday market and a fair in the same year; Kirkby Malzeard gaining a Monday market and two fairs three years later, while Pateley Bridge obtained its Tuesday market and two fairs in 1320. Given the nature of the competition, the market at Ripley did reasonably well, surviving into the latter part of the eighteenth century (McCutcheon, p. 176). From the time that the market was gained it is safe to assume that the lords of Ripley will have been considering how best to locate it, attract custom and integrate it with the affairs of village and estate.

In 1360, Sir Thomas is said to have acquired the neighbouring manor at Nidd, adding it to the several estates in his possession (Lancaster, p. 14, citing Fines, 34 Edw. III). This would explain the road from Ripley to Nidd via the marl quarries which slices across the furlongs and plough strips on both the Nidd and Ripley sides of the township/parish boundary. To create a road with such a disregard for established farming patterns on two sides of a boundary would require special agreements between the respective lords unless the land concerned was in the ownership of a single lord. The late medieval roads that were created to reach the Killinghall bridge on the river Nidd and Nidd village appear to have dismembered a large furlong recorded, in 1752 as 'Aunams', and perhaps in 1202 as 'Elum'. The name 'Aunams' seems to be a meaningless corruption of an earlier name, which appears to have been Elum, perhaps signifying elms, alders, elders, land growing flax, but quite possibly meaning something else. The furlong was mentioned in 1202 when Thomas son of Serlo released land in Ripley to Richard de Rippley, ... 'who granted a parcel of land in Elum, in Ripley, to the same Thomas ...' (Farrer, 1914, notes to charter 524, p. 405). In 1276 the demesne of the larger manor was said to include selions in the meadows or furlongs called Sourlands, Elgilsteills, Greteholmes, Braceynges and two selions in Bank of Elum next to the land of the church (Ingilby MS 188).

There still remains the question of why farming arrangements should have been disrupted in the two townships when an apparently serviceable lane already connected the two manors? This was the 'northern route', running from Knaresborough through the medieval village of Nidd (removed in stages during the nineteenth century in the course of park-making), past Nidd church and then westwards as Nidd Lane to join the Skarah–Ripon routeway. It might well be that the Ingilby concerned was seeking sole control of the link to avoid any dependency upon his neighbour in Ripley township, the Abbot of Fountains. The date when the new roads were driven across the old ploughland is uncertain, but a southern outlet leading to the bridge and Nidd roads was built into the lay-out of the planned medieval Ripley village.

The church

When the first Sir Thomas and his wife died they were buried in the old church overlooking the River Nidd. (Ripley lost its lord and its rector, Joseph Mauleverer, in 1369, one or both perhaps the victims of one of the three recurrences of the Pestilence that took place in this decade). The family tree is debated, but it seems that Thomas and Edeline's eldest son was another Thomas. He lived from about 1325 to 1393, had a wife, Ellinora, and was the founder of the current Ripley church. Thorpe, who seems to have had access to interesting sources that he never properly referenced, wrote that Richard Kendall was the last rector of the old church and the first rector of its successor and that he died on 4 January, 1429. Henry de Ingilby became rector in 1382, and so the dating evidence suggests that the new church would have been built in the years around 1390. The arguments concerning the role of environmental forces in undermining the old church have already been aired. It

FIGURE 37.
The church at Ripley, still largely of the late fourteenth century.

PLATE 1 (*facing page*). The Ilkley–Aldborough Roman road. Off the map to the west, the road traverses the watershed plateau from Wharfedale to Nidderdale, and as it begins to descend into the valley of the Nidd its course is marked by the A59. At Kettlesing Head both modern and Roman roads followed a contour curving round to the ESE, but in the vicinity of the Black Bull Hotel, the two roads diverge, with the Roman route heading towards a fording place in the vicinity of Hampsthwaite. At first, its course appears to vanish in the pastures as it drops to the valley bottom.

Between A and the river bridge at Hampsthwaite, the Roman road appears to form the spine of the old village, which is plainly aligned upon a track leading to the river crossing. In 1999, excavations by Kevin Cale in the vicinity of B produced evidence of an old paved track, which could be medieval.

The route crosses the broad floodplain of the Nidd at C, probably on the line adopted by the minor Hampsthwaite–Clint road. The road can be seen to be slightly causewayed. Having traversed the floodplain, it curves up the steep slope of the valley side.

The lane to Clint and the Roman route then diverge. The Romans were seeking a direct route to Aldborough, but having crossed the Nidd their choice of a line was partly governed be the meander loop which existed in Roman and medieval times and which would rule out more southerly courses. In the region CDE and beyond towards Ripley, the Roman road was adopted by the old, narrow lane that was formerly the main road between York and Lancaster.

At D a marked 'wiggle' in this lane must mark a brief diversion from the straighter Roman route. Buildings may have obstructed the original track, but it is also possible that the wiggle represented an adjustment to the introduction of a new little system of fields in Clint.

At E there are problems: the bend to the NNE seems unusually sharp for a Roman route, but if a more direct route was employed then it has disappeared without obvious traces. One possible reason for the diversion is the steep, west-facing old river bluff in Hollybank Wood. Were heavily-laden packhorses or carts using this road, then the deviation to a gentler gradient might be justified.

Between F and G the alignment of the Roman road along Hollybank Lane is marked by the massive grit kerbstones marking out the curving course of the Roman route.

At G, and close to the medieval moat at Sadler Carr, the Roman track seems to be completely lost in the field, with a section heading to the Ripley Beck at H being buried under landfill materials. A later track diverts towards Ripley at G, perhaps being a medieval route approaching the manor/village via a bridge provided by a convenient mill dam.

As soon as the Ripley Beck is crossed, the position of the Roman route seems to be marked by a short floodplain causeway and the holloway that ascends the valley slope.

In the vicinity of J the Roman road was preserved for centuries as the boundary and access-way between two furlongs.

At K the Roman route is faintly recognisable as a track on the south side of an old hedgerow; this was marked as a lane on the Chippendale survey of 1752.

Two possible routeways must be considered at L. A short holloway running down to the Beck marks a more southerly possibility – and certainly one that was used at one time or another. the more northerly route, which swings sharply NE before crossing the stream, is more closely attuned to the medieval field boundaries. Earthwork evidence to the east has been destroyed in the area where medieval and later quarrying was followed by the construction of a sunken caravan park.

The track then becomes plainly apparent as it runs to the south of ancient hedgerows. At N it forms the boundary between the parishes of Ripley and Nidd and is preserved as a substantial flat-topped, ditch-girt bank. Nidd church was apparently positoned to be beside this Roman routeway, and although the line briefly vanishes in Nidd park, it soon re-emerges and can be traced to Aldborough.

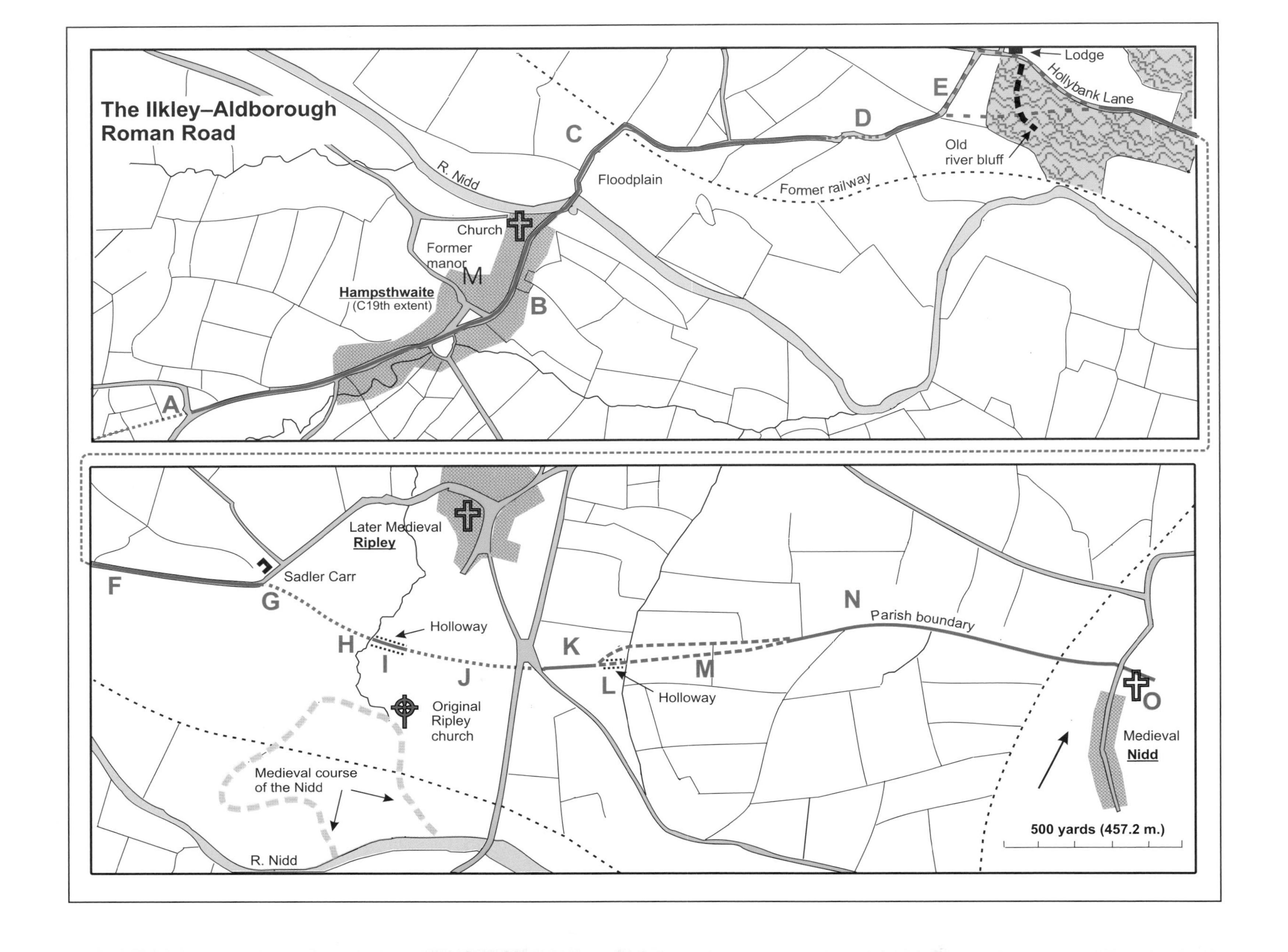
The Ilkley–Aldborough Roman Road
Lodge
Hollybank Lane
E
D
C
Old river bluff
Former railway
Floodplain
R. Nidd
Church
Former manor
M
Hampsthwaite
(C19th extent)
B
A
Later Medieval
Ripley
F
Sadler Carr
G
Holloway
H
I
J
K
N
Parish boundary
M
L
Holloway
Original Ripley church
O
Medieval
Nidd
Medieval course of the Nidd
R. Nidd
500 yards (457.2 m.)

PLATE 2 (*left*).
One of several house platforms beside the Roman road in Hollybank Wood.

PLATE 3 *(below)*.
The view looking back across the river from Killinghall. The original church lay amongst springs on a steep river bluff in the woods at the centre of the photograph.

PLATE 4 (*facing page*).
The Sinking Chapel locale, Ripley.

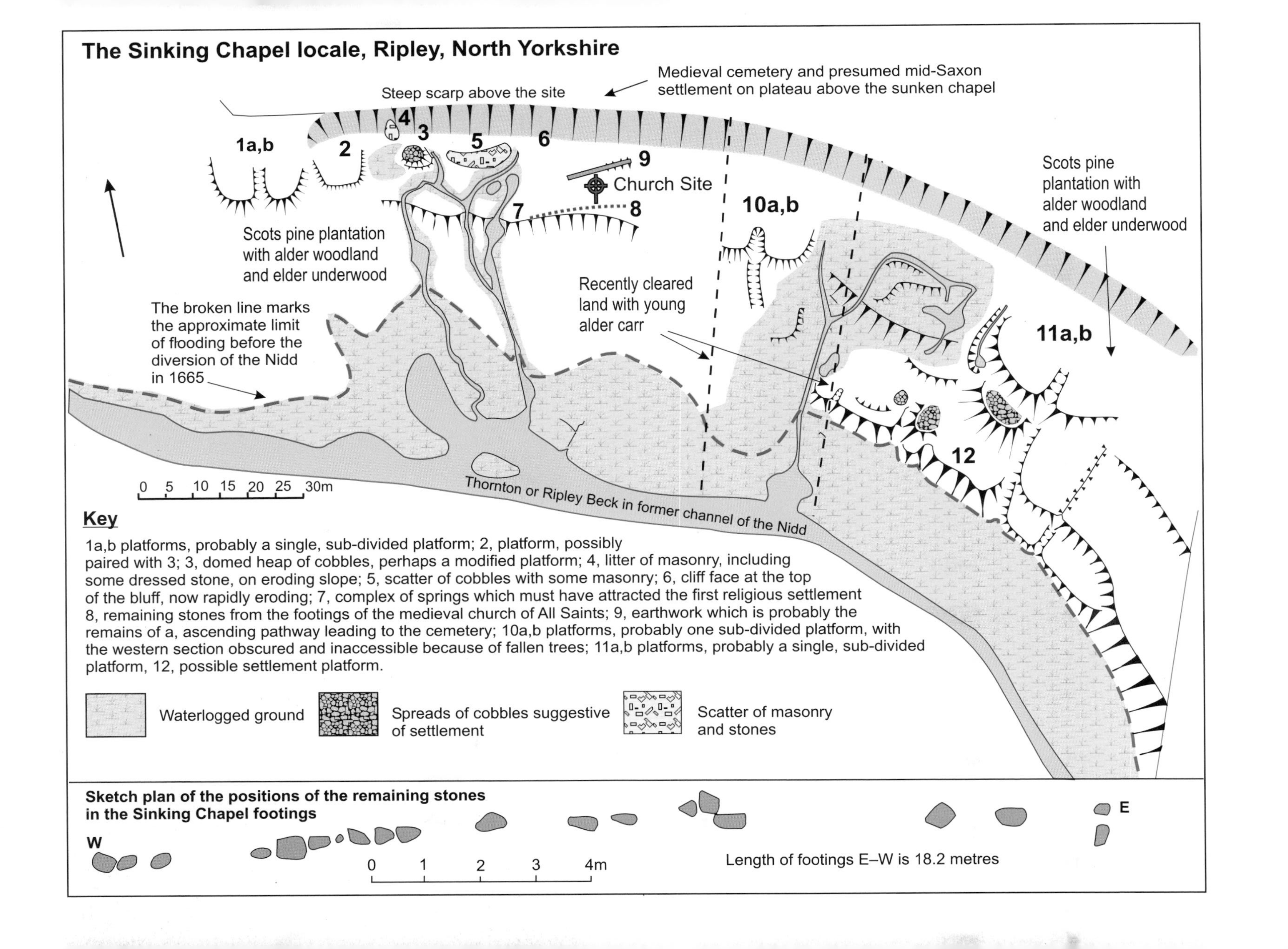
The Sinking Chapel locale, Ripley, North Yorkshire
Steep scarp above the site
Medieval cemetery and presumed mid-Saxon settlement on plateau above the sunken chapel
1a,b
2
4
3
5
6
9
Church Site
8
7
10a,b
11a,b
12
Scots pine plantation with alder woodland and elder underwood
Scots pine plantation with alder woodland and elder underwood
Recently cleared land with young alder carr
The broken line marks the approximate limit of flooding before the diversion of the Nidd in 1665
Thornton or Ripley Beck in former channel of the Nidd
0 5 10 15 20 25 30m
Key
1a,b platforms, probably a single, sub-divided platform; 2, platform, possibly paired with 3; 3, domed heap of cobbles, perhaps a modified platform; 4, litter of masonry, including some dressed stone, on eroding slope; 5, scatter of cobbles with some masonry; 6, cliff face at the top of the bluff, now rapidly eroding; 7, complex of springs which must have attracted the first religious settlement 8, remaining stones from the footings of the medieval church of All Saints; 9, earthwork which is probably the remains of a, ascending pathway leading to the cemetery; 10a,b platforms, probably one sub-divided platform, with the western section obscured and inaccessible because of fallen trees; 11a,b platforms, probably a single, sub-divided platform, 12, possible settlement platform.
Waterlogged ground
Spreads of cobbles suggestive of settlement
Scatter of masonry and stones
Sketch plan of the positions of the remaining stones in the Sinking Chapel footings
W
E
0 1 2 3 4m
Length of footings E–W is 18.2 metres

PLATE 5 (*left*).
The dawn light displays the outline of a former building on the margins of a former furlong in a field between the village and the old church.

PLATE 6 (*below*).
Curving medieval cultivation terraces surviving as earthworks beside the Ripley Beck.

PLATE 7 (*facing page*).
Ripley: the environment in the twelfth and thirteenth centuries.

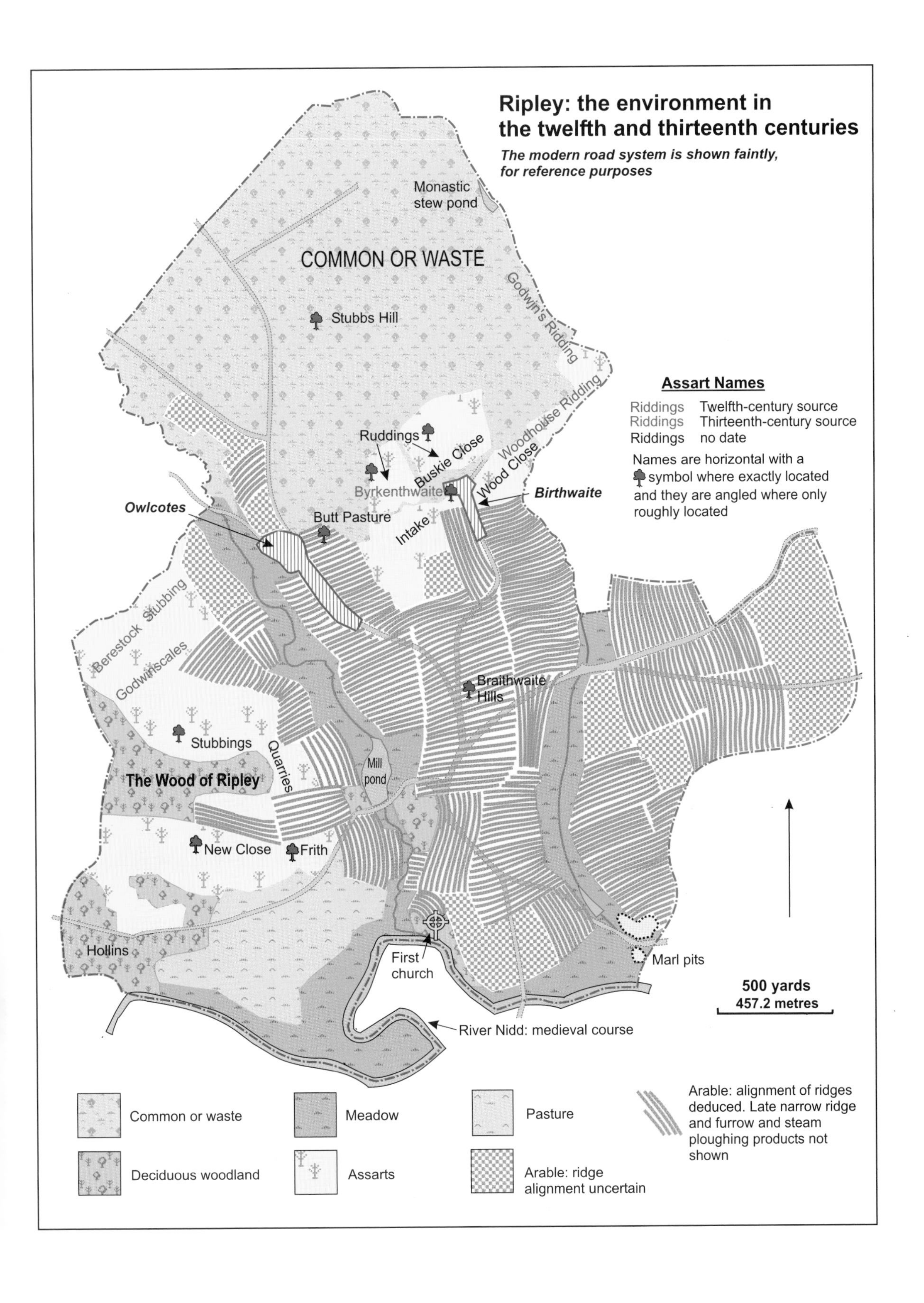

Ripley: the environment in the twelfth and thirteenth centuries
The modern road system is shown faintly, for reference purposes
Monastic stew pond
COMMON OR WASTE
Stubbs Hill
Godwin's Ridding
Woodhouse Ridding
Ruddings
Buskie Close
Wood Close
Byrkenthwaite
Birthwaite
Owlcotes
Butt Pasture
Intake
Berestock Stubbing
Godwinscales
Braithwaite Hills
Stubbings
Quarries
Mill pond
The Wood of Ripley
New Close
Frith
Hollins
First church
Marl pits
River Nidd: medieval course
500 yards
457.2 metres
Assart Names
Riddings Twelfth-century source
Riddings Thirteenth-century source
Riddings no date
Names are horizontal with a symbol where exactly located and they are angled where only roughly located
Common or waste
Meadow
Pasture
Arable: alignment of ridges deduced. Late narrow ridge and furrow and steam ploughing products not shown
Deciduous woodland
Assarts
Arable: ridge alignment uncertain

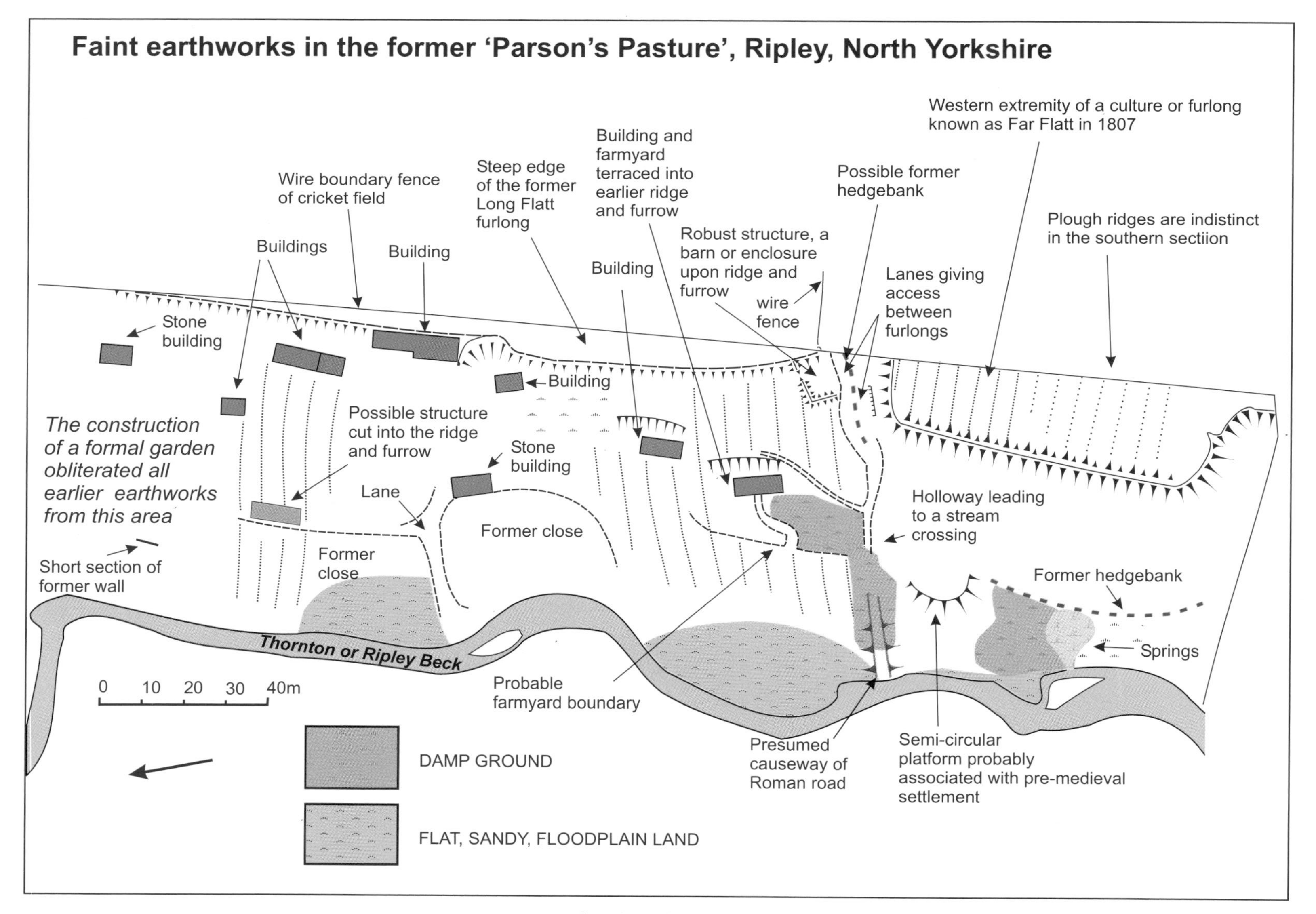

PLATE 8. Faint earthworks in the former 'Parson's Pasture'.

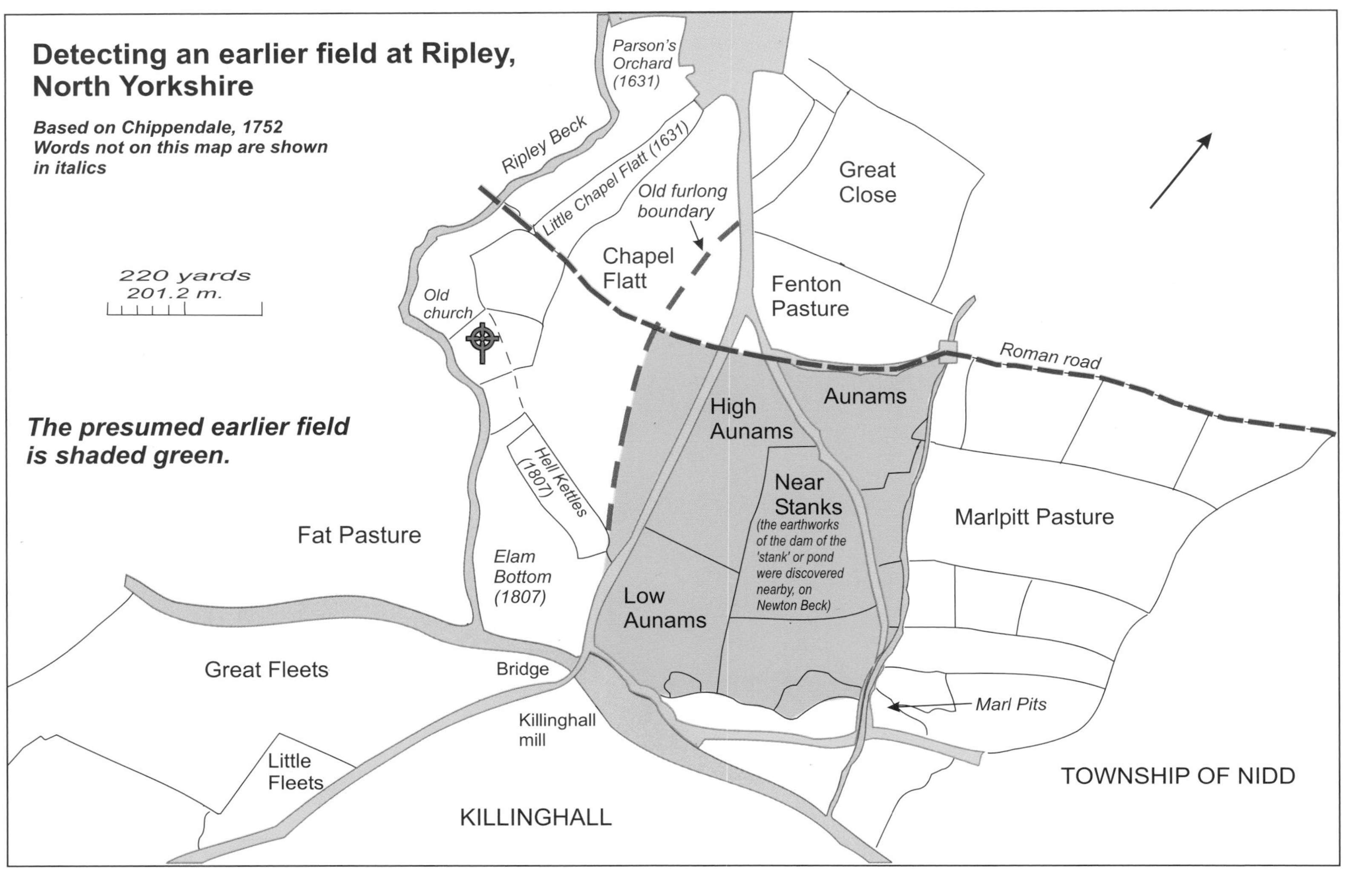

PLATE 9. Detecting an earlier field at Ripley. The Aunams or Aurams field names, the configuration of the field patterns as mapped in 1752, the co-axial field network and the later, superimposed character of the roads suggest the presence of an earlier large, rectangular unit. The name is open to various translations and could be a reference to the golden colour of the grain crop. There are medieval references for a field called 'Elum', probably preserved in 'Elam Bottom' and this might have been the predecessor of Aurams.

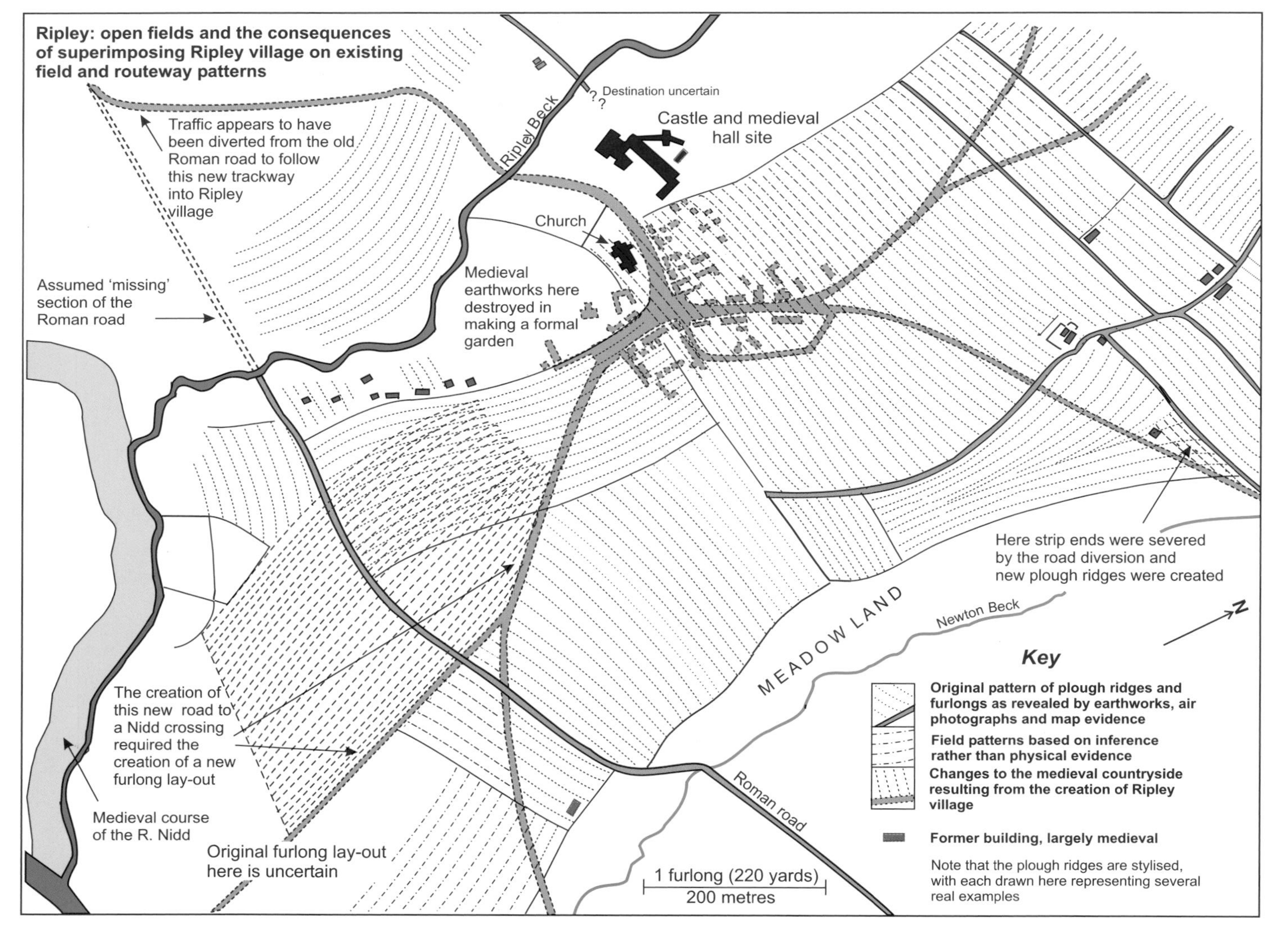

PLATE 10. Ripley: open fields and the consequences of superimposing Ripley village on existing field and routeway patterns. Roads to the north seem to have been 'pulled' in to the new village, though their old courses seem to have vanished from the landscape. While the village and church appear to have been planted on open field plough ridges, the older manor house seems to have been slightly to the west of the ploughland, where the castle now stands.

PLATE 11.
Heavy horses grazing on plough ridges severed by the making of roads running southwards from the new village.

PLATE 12.
Ripley Church seen from the east.

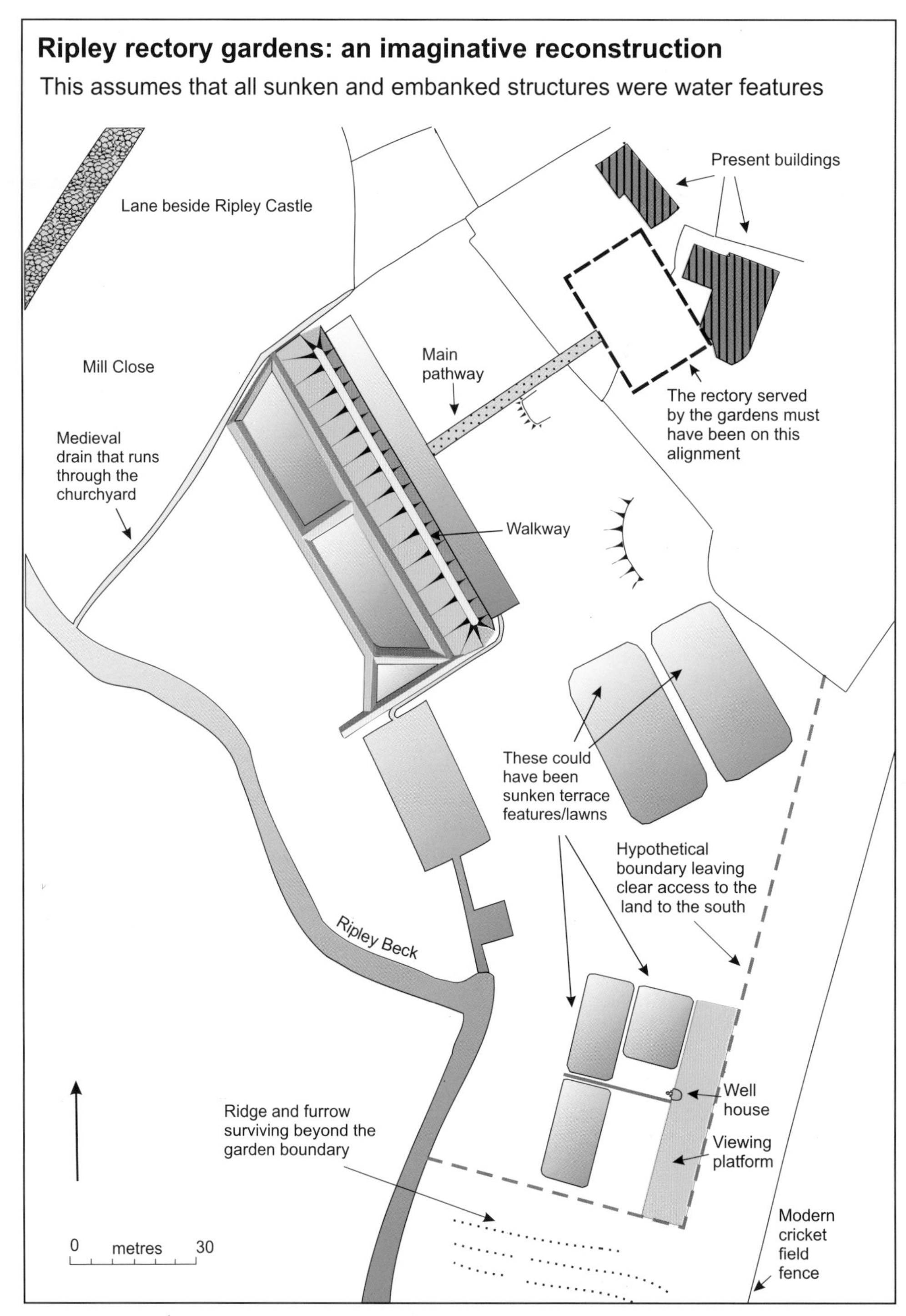

PLATE 13. Ripley rectory gardens: an imaginative reconstruction. The form of the main feature seems to have been determined by the old drain that borders the garden. Seventeenth-century rectors were punished by the manor court for failing in their responsibilities for cleaning and scouring it. The precise position of the rectory that overlooked the garden is not known. The building existing at the time of the 1672 hearth tax was still a relatively modest one, with 5 hearths compared with the 24 at the nearby castle.

PLATE 14.
The sixteenth-century tower (*centre left*) and more recent additions at Ripley Castle.

PLATE 15.
Sheep graze on the earthworks of the formal garden.

PLATE 16.
Landscaping from the mid-nineteenth century in the late medieval deer park.

PLATE 17.
Given the quality of its village landscape it is not surprising that Ripley attracts hundred of visitors each weekend during the tourist season.

FIGURE 38.
The rood screen, fourteenth-century in style, which appears to have been removed from the original church, shortened slightly, and inserted into the new church.

FIGURE 39.
The nave of the Ingilbys' church at Ripley.

is hard to imagine that the establishment of the Ingilby family in Ripley and the decision to relocate what may well have been a fairly new and perfectly serviceable church were unrelated. By moving and rebuilding the religious focus of the parish they could stamp their authority upon the landscape in a manner that none could ignore. Equally, the church relocation could have been a key component in a greater strategy for reorganising the fundamental structure of the township.

The church was moved in an almost literal sense. The first Sir Thomas and his wife were disinterred and reburied when their tomb chest was installed in the new church. The rood screen was transferred, the weeping cross was probably moved and it seems, from the relative lack of masonry at the original site that the stones from this church were transported and reused in the successor. The site chosen for the replacement appears to have faced the manor house across the lane which led over the mill dam towards the Roman route at Sadler Carr and lay by the junction of the core open field ploughland and the meadows flanking the Ripley Beck. The church built here at the end of the fourteenth century is essentially the building that is seen today. A south aisle was added in the fifteenth century, while in the 1560s, the roof of the nave was raised to accommodate a clerestorey and the tower was heightened. With little regard for the sanctity of the original graveyard, stones were removed to provide lintels for the extensions to the new church. During the eighteenth century, the roof was lowered, the floor was raised and pews were introduced, while the cylindrical piers were encased in plaster and made to appear octagonal with square capitals, though some of the modifications of the previous century were removed in the 1860s.

Inside the church the monuments transferred from the 'Sinking Chapel' include the tomb chest with its effigy of an armoured knight and his lady, with eight 'weepers', six of them representing their children, surrounding the chest. Beside the tomb is a cut-down tomb slab of a fourteenth-century priest inscribed with a cross, book and chalice. This could belong to Joseph Mauleverer, who may also have been exhumed. (The grave slab of Richard Kendale, priest at both churches, is in the chancel beside the vestry door.) A gravestone from the old church decorated with a fourteenth-century ceremonial water vessel or 'laver' is at the end of the nave to the north of the founder's tomb. In the churchyard, and also reputedly transported from the old church is the unusual cross base known as the 'Weeping Cross'. Eight recesses are cut in the lower of the two limestone cylinders, presumably to house the knees of kneeling penitents, while the upper one has a socket which will originally have held the cross shaft. This monument appears to be a unique survival, though the

FIGURE 40. This fourteenth-century style gravestone engraved with a ceremonial vessel for water is preserved in Ripley Church along with the slabs of two early priests. It may be that of a priest who was buried in the original church.

FIGURE 41. This extremely unusual penitential cross base is thought to have been brought to the new church at Ripley from its predecessor.

biblical event or the festival associated with its use have been forgotten. Apparently ceremonial expressions of grief or penitence were at the core of the ritual.

The village

That a medieval model village of Ripley was created in the decades around the start of the fifteenth century is beyond dispute. The precise dating of the foundation is debatable, with a number of different dates and founders being available for evaluation. In situations such as this the research process can be clarified by tabulating the facts and proceeding to evaluate the different possibilities.

Date	*Founder*	*Occasion*
1357	Sir Thomas Ingilby (1)	Coinciding with the acquisition of a market charter
c. 1390	Sir Thomas Ingilby (2)	Coinciding with the building of the new church
c. 1414	Sir Thomas Ingilby (3)	Coinciding with consolidating the demesne

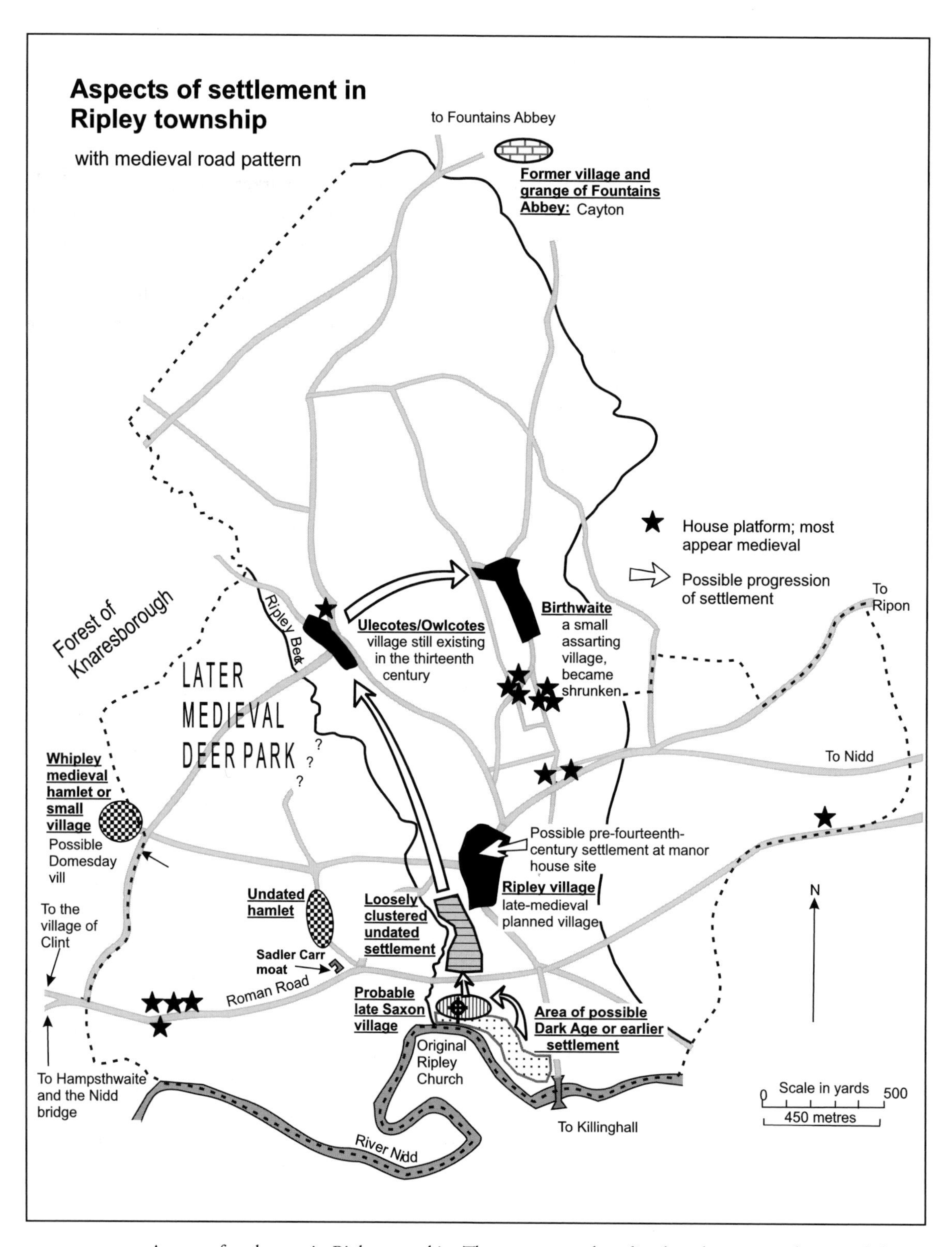

FIGURE 42. Aspects of settlement in Ripley township. The reconstructed medieval road pattern is shown and the arrows indicate the possible clustered settlement succession.

Each possibility has considerable attractions. With regard to the first option, the gaining of a market charter seems to have been the initial step in the establishment of a number of planned medieval settlements. As we have noted, having obtained the right to hold a market in 1357 Sir Thomas Ingilby would certainly have needed a suitable venue for trading. While a few of the great annual fairs were held in rural locations, weekly markets were almost invariably staged on the market greens or squares of a town or village. The charter provides the best remaining evidence for the existence of a village of Ripley in the mid-fourteenth century.

The fact that it was possible, around 1390 to establish a church in the place that was/became the village core argues against the prior existence of a village – for had the village already existed then the core space would have been occupied and the church would have had a more peripheral situation. One could imagine a scenario in which the 1357 market was located beside the Ingilby manor house and its accompanying hamlet. With the commercial development of this locality, the old church and its setting might have seemed remote and marginal, leading to a decision to relocate the church in the manor/market nucleus. Alternatively, as I have just noted, the church appears to be a key component of the village lay-out rather than a later addition, and so it could be argued that church and village were crucial elements in an integrated plan for reorganising the cultural landscape.

Thirdly, it might be argued that documented events early in the fifteenth century argue for a slightly later date. In 1411 John Wilkinson swopped seven acres of land at Hallsteads in the demesne of Ripley with Sir Thomas Ingilby for a close (a field or enclosure) called 'Bilhous' – 'Bill's house' (Ingilby MS 201). In 1412, John and Joanna Mees of Clint agreed, for the sum of £40, not to sell or alienate land in Broxholme or Whipley except to Sir Thomas (Ingilby MS 203). In the following year the same parties agreed on the granting to Sir Thomas of half an acre in Hallsteads (Ingilby MS 204) and in 1414 Thomas Vasavour gave Sir Thomas two acres in Hallsteads and half an acre and half an rood in a field called Fowlowe in the demesne of Ripley in exchange for land in the east of the township (Ingilby MS 206). Plainly, Sir Thomas was seeking to consolidate the demesne lands at Ripley and to obtain a compact estate to facilitate farming. Such activities were frequently pursued amongst the lords of medieval manors, but might not Sir Thomas have had a more specialised purpose? With the exception of the Broxholme and Whipley territories the lands involved were in Hallsteads, demesne land lying close to the manor house and village. The planned late-medieval village of Ripley trespassed on the communal ploughland (Plate 10). Some of it may well have belonged to the Ingilby family, but other selions might have had to be bought or obtained by land swaps to free the area designated for development.

Any of the three factors mentioned could have been the trigger or the key to village formation, though one of the dates in the fourteenth century is probably preferable to the later one.

Ripley was a precisely planned village, and since medieval buildings were most probably still standing in 1752 when the first detailed and accurate map was drawn, we can be sure of its morphology or lay-out. The plan adopted was a variation on the locally popular 'Y'-shaped model exhibited in the township's older nucleations at Birthwaite and, to some extent, Owlcotes. The stem of the 'Y' was formed by the lane arriving from across the mill dam and was flanked to the north by the castle and to the south by the churchyard. The space in the angle of the 'Y' was filled by the market and the limbs of the 'Y', linked by a head row at the top of the market, were lined by dwellings. The form of the road running through the southern limb showed that the route to Killinghall bridge and the 'southern' route to Nidd were integral to the lay-out, while a lane may have run to the old churchyard.

Clear and detailed topographical information is seldom evident in medieval documents so that the following description is better than one might reasonably expect: 'Lease from Richard de Kendale, rector of the parish church of Ripley, with the consent of William de Ingilby, son and heir of Thomas de Ingilby, to John Wymmerlee of Ripley, of a messuage (farmstead) with a cottage, between Richard Sklater's messuage and Robert Barker's messuage, a croft, 2 bovates and 3 roods land and meadow, and a cottage next the cemetery and opposite the messuage in Ripley, for 30 years at 20*s.* 8*d.* (£1.03) annual rent. Wymmerlee to build a new messuage ... 14 July 1426' (Ingilby MS 214). This clearly contains a description of part of the village and tells us that by 1426 the place was very much in existence. Secondly, as well as mentioning a cottage standing next to the cemetery, it shows that farmsteads standing side by side lined part of the street facing the cemetery. The third point is more curious: if the lessee, John Wymmerlee, was required to build a *new* farmstead, how long had the village existed? The hovels of peasants living in earlier medieval centuries had only endured for around a generation before requiring a rebuilding, but late medieval farmsteads would be better built than that. If Wymmerlee was to build an additional, an enlarged or an alternative farmstead then the passage has no special significance, but if the building concerned was already dilapidated and worn-out, then the village should have existed for quite a few decades, favouring the earliest of the three dates, the market date.

The park

There is no doubt that a deer park existed at Ripley in Tudor times, though some mysteries attach to its existence. For although the creation was exceptionally late for a medieval park of this kind, the documentation concerning its establishment is remarkably sparse. Domesday Book records 35 deer parks, the heyday of the parks coming after the Norman introduction of fallow deer, with around 3,200 examples existing around 1300, when they covered about two per cent of England (Rackham, 1994, p. 59). The Ripley park may not have come into existence for well over a century after the heyday of the deer park in England. It might be argued that the land exchanges of the early fifteenth century, mentioned above, represented a move to control directly the lands designated for the park (Jennings, p. 108). However, it would be strange for a further three-quarters of a century to elapse before the deer park found its way into the records. Even this mention from 1488 (Ingilby MS 227) is unconvincing, being a grant of land that mentions a rood of land in le Pacrofte between the land of the Abbot of Fountains from the south and the land of John Banarr from the north. 'Pacrofte' might be an abbreviation of Park Croft, but the terminology and mention of a headland are more suggestive of open field ploughland. It is only in the sixteenth century that mentions of a park become frequent. In 1523 Sir William Ingleby leased to the rector, Thomas Skawesby, a wooded close adjoining the mill beck and lying between the parson's orchard and the park called 'Mill Close', and in return he leased a close in his own park called Tyndall Close. The rector also leased Barkhouse Garth, lying between the mill beck and the park, while allowing Sir William access to the two closes with his carriages (Ingilby MS 1064). Two years later, Thomas Beckwith of Clint renounced his own and his tenants' rights to use the common highway running through the park from Clint township to Ripley church and to the markets at Knaresborough, Ripon and Boroughbridge (Ingilby MS 1065). From this evidence it seems that the legal arrangements and land exchanges associated with consolidating, if not actually founding the park were taking place in the 1520s.

There is, however, one piece of evidence that suggests a significantly earlier dating for the park. Sir John Ingilby was born in 1434 and inherited his estates when he was only aged five. During the troubled reign of Henry VI, corroboration of his date of birth was required before his inheritance could be returned to him from his trusties. Witnesses were required to testify to his date of birth: 'Ralph Acclom remembers John's birth because he was staying with John, Abbot of Fountains Abbey and rode across with him to baptise the baby. Robert Apilton remembered John's birth because he killed a deer between Ripley and Hampsthwaite . . .' (Ingilby, p. 4). The area of the deer park extended for about half of the way from Ripley to Hampsthwaite. If Robert Apilton had shot his deer in the farmland outside Ripley township then he could only have been in the Forest of Knaresborough – and he would have been unlikely to admit to shooting one of the King's deer. If the park had been established by 1423 then it could more easily be regarded as a part of a comprehensive church/village/park and roads project of rural transformation.

Former generations had different ideas about the meanings of 'new' and 'old'. Thus the reference in 1547 to Sir William's 'newly built' park in Ripley in a grant of land called 'Reyst Close' inside the park by Richard Atkinson of Killinghall to Sir William might be stretching the meaning of 'new' – or else be referring to a recent extension of the park. It is often imagined that within its corset of banks, walls, palings or ditches the deer park was a stable unit of land. This was frequently not the case, and Ripley park has certainly fluctuated considerably in its extent throughout the centuries.

There is no doubt that its acreage was increased at a fairly early stage in its existence. In the 1570s a dispute arose between Thomas Beckwith of Clint and the churchwardens of Ripley concerning the financing of repairs to the church, when Beckwith alleged that Sir William Ingilby should pay dues to Clint rather than to Ripley for the '*old and new* parks' of Ripley (Ingilby MS 2985). Perhaps the strangest aspect of the puzzle concerning the dating of the park is the absence of a royal licence allowing a park to be created at Ripley. Such licenses to empark were a useful source of revenue to the Crown, and were considered essential for any park, like the one at Ripley, that bordered upon a Forest

> Deer were considered the property of the Crown, and in theory a royal license to embark was requested. The large number of well-documented parks for which no license appears to have been granted suggests, however, that one was not necessarily sought, except where a proposed park lay close to, or within a royal forest. Grants of license to empark were occasionally made by the Crown as a means of rewarding an official for services rendered, or simply as a way of indicating royal favour (Neave, 1991, p. 5)

Since the arrival of a deer park at Ripley came so late in the medieval period, one might wonder whether the early sixteenth-century park was more an ornamental creation than a hunting reserve? In fact that there is good evidence that hunting did take place there well after the close of the medieval period. The almanacs of Sir William Ingilby include the following entries:

> 1 Sept 1665: 'I killed a Blacke Bucke out of the Parke at Rypley ... another ? before 3 fawnes this year count ten is all left ... 2 bucks killed out of the paddock this year'.

and

> August 24 1668: 'I killed a fat Buck out of the padock ... 7 faunes'. Ingilby MS 3590.

The archaeology of the park also produces evidence of hunting at a slightly earlier stage, perhaps from its creation around 1500. Near the highest ground there is a complex of monuments, including a paved but overgrown trackway and some settlement traces. Curiously separated from the holloway of the paved route by a parallel holloway-like feature is a shallow ditched mound somewhat resembling a small, very degraded motte mound or round barrow. The mound or platform incorporated earth and stones, many of which are reddened by burning. It carried a rectangular stone structure measuring 8x5 metres, the footings of which can be traced, and this is regarded as a hunting tower (S. Moorhouse, pers. comm.). The host and his guests would have been stationed in the tower to shoot at game driven past their retreat by the servants. Afterwards, the men may have spent the evening carousing in the tower. There is also evidence from field names that hunting dogs were kept. On a map of 1807 two closes named 'Dog Croft' are shown in the park between Broxholme and Whipley, close to where a hunting lodge would be expected, while the

FIGURE 43.
Hunting tower at Ripley: general context.

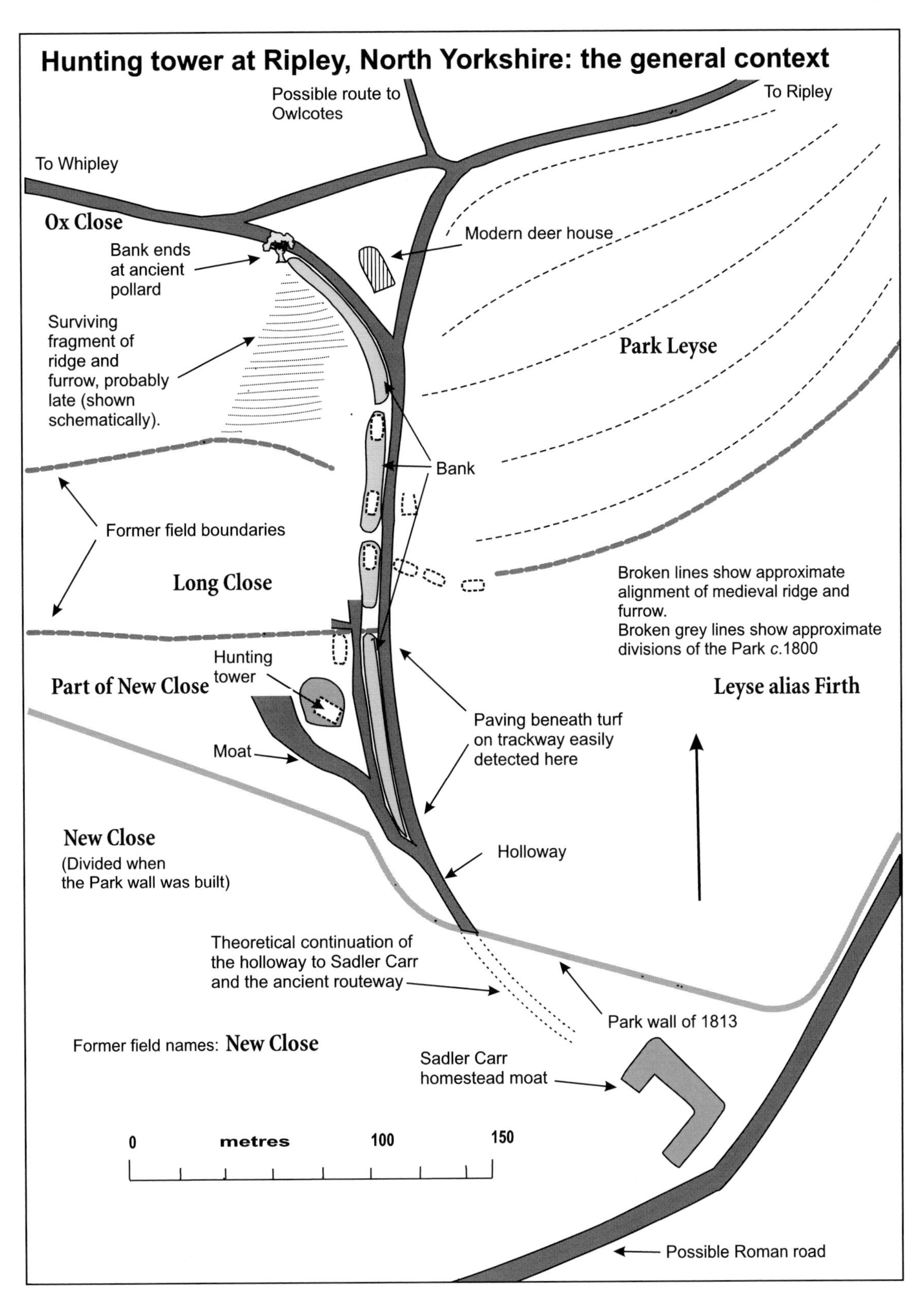
Hunting tower at Ripley, North Yorkshire: the general context
Possible route to Owlcotes
To Ripley
To Whipley
Ox Close
Bank ends at ancient pollard
Modern deer house
Surviving fragment of ridge and furrow, probably late (shown schematically).
Park Leyse
Bank
Former field boundaries
Long Close
Broken lines show approximate alignment of medieval ridge and furrow.
Broken grey lines show approximate divisions of the Park c.1800
Hunting tower
Part of New Close
Leyse alias Firth
Paving beneath turf on trackway easily detected here
Moat
New Close
(Divided when the Park wall was built)
Holloway
Theoretical continuation of the holloway to Sadler Carr and the ancient routeway
Park wall of 1813
Former field names: New Close
Sadler Carr homestead moat
0
metres
100
150
Possible Roman road

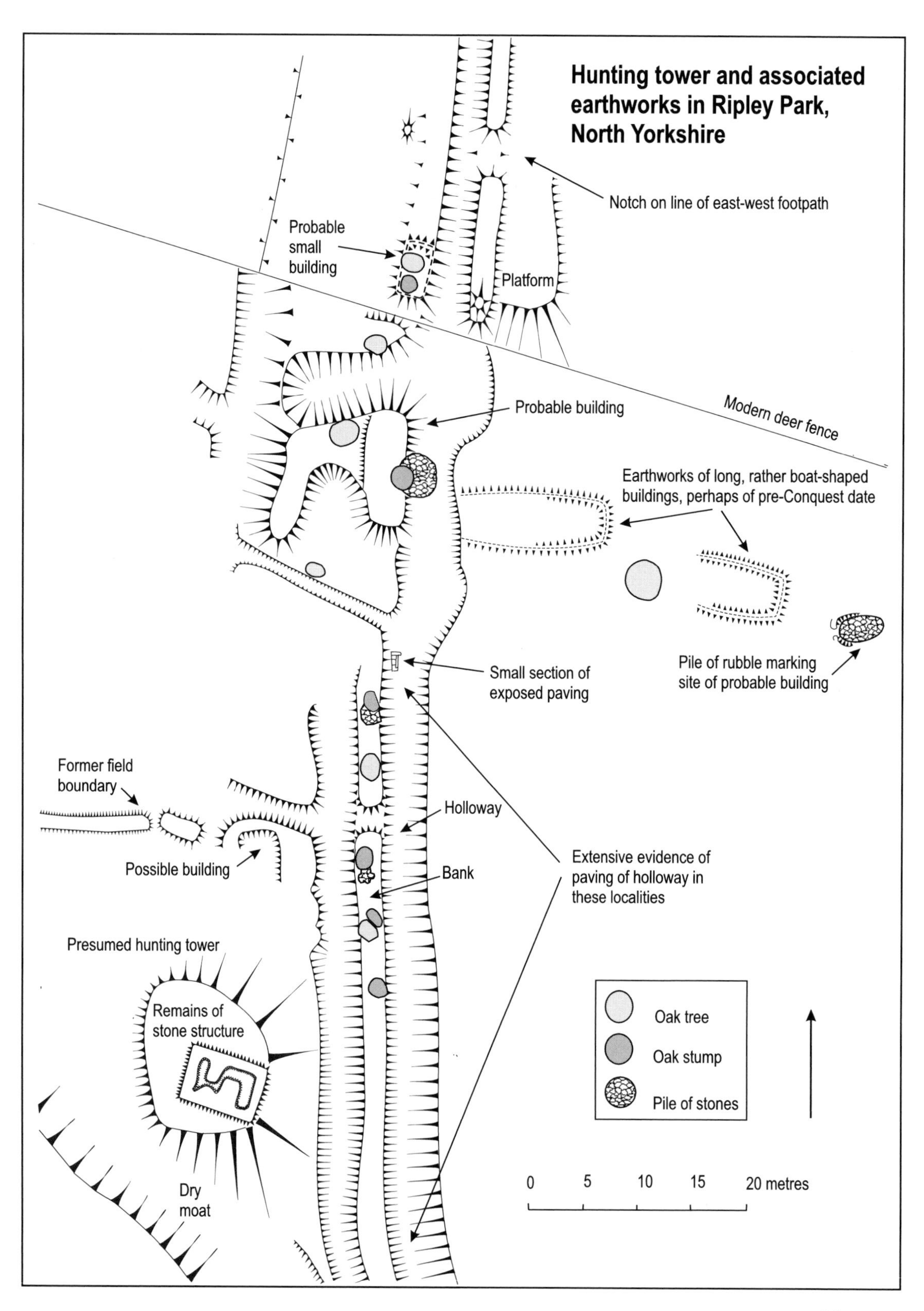
Hunting tower and associated earthworks in Ripley Park, North Yorkshire
Notch on line of east-west footpath
Probable small building
Platform
Probable building
Modern deer fence
Earthworks of long, rather boat-shaped buildings, perhaps of pre-Conquest date
Small section of exposed paving
Pile of rubble marking site of probable building
Former field boundary
Holloway
Possible building
Bank
Extensive evidence of paving of holloway in these localities
Presumed hunting tower
Remains of stone structure
Oak tree
Oak stump
Pile of stones
0
5
10
15
20 metres
Dry moat

FIGURE 44 (*facing page*). Hunting tower and associated earthworks in Ripley Park. The purpose of the 'double holloway' in the southern section is puzzling. On the intervening bank low heaps of stone might derive from tumbled walls, though some could well be associated with former buildings.

monastic right of way running down from here just outside the pale to the park gate on Hollybank Lane, known as Dob Lane or Dog Lane, might have been used for exercising the dogs.

Northern deer parks were not the simple, exclusive reserves of popular imagination, and hunting was normally just one amongst a number of activities pursued within them. In the case of Ripley one or more large paddocks seem to have been set aside for deer, while fields surviving within the boundaries were apparently tenanted and worked for agricultural purposes. Equally, parts of the park could be excluded and returned to agriculture, while even a building as central to hunting as the lodge itself might be rented-out. In 1684 Sir John and Lady Ingilby leased John Hare and Nicholas Bradley: '... a messuage in Ripley Park, called the Lodge, and parcels of ground called Low Forth, Stelling, High Forth, the Old Park and Park Lees for 11 years at £5 ...' (Ingilby MS 2125).

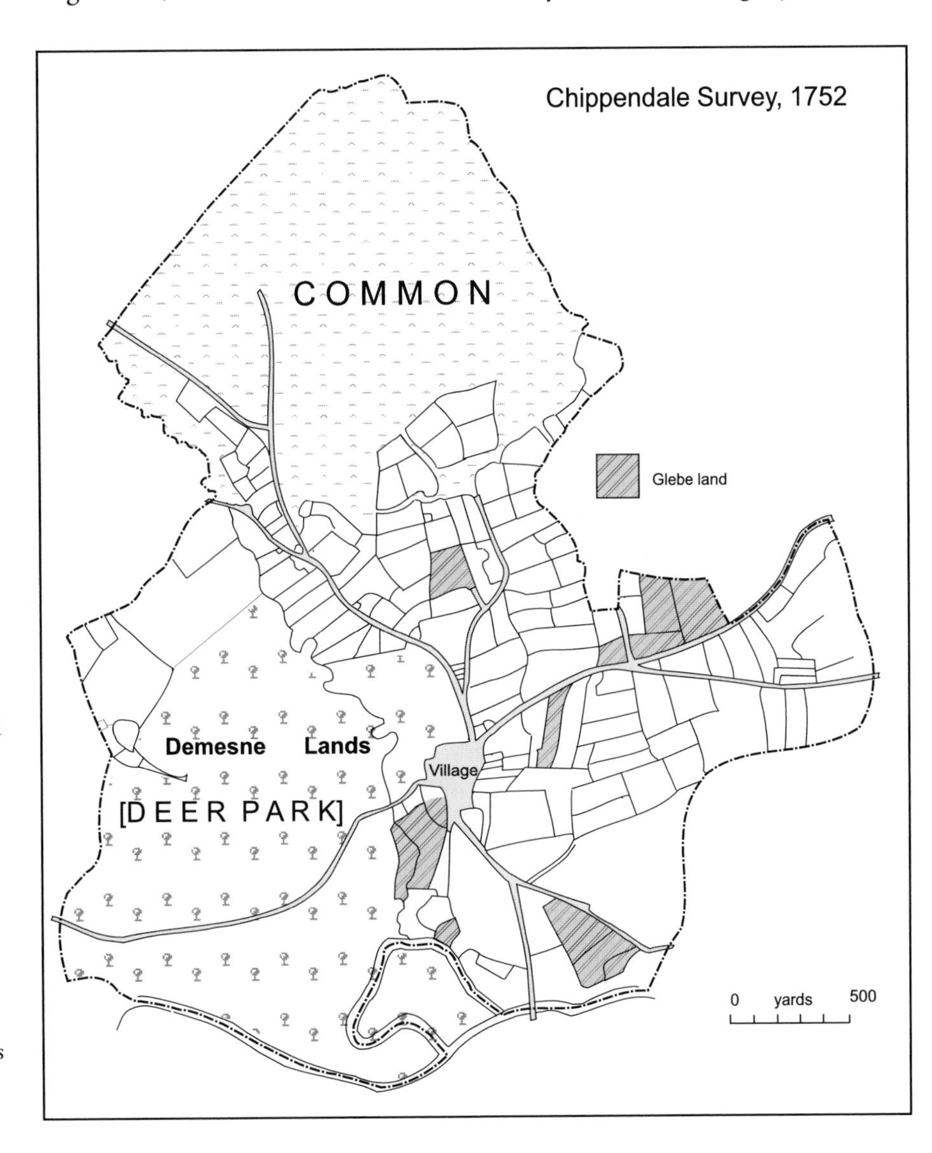

FIGURE 45. The Chippendale Survey, 1752. Undertaken by a local man who was related to the celebrated designer of furniture, the survey provides the earliest surviving cartographical representation of the township. The areas with the diagonal shading are glebe land. The deer park is at almost its greatest extent.

This did not signify the end of the park, just a stage in the fluctuation of its boundaries. By 1752, with the loss of some land on its northern margins, the park was essentially the medieval territory of Godwinscales, but by 1807 it had retreated to a fraction of this extent, though it would expand as a park landscaped in the manner of Lancelot 'Capability' Brown later in the century. Ripley park is notable as one of the few medieval deer parks still to contain deer. About 80 fallow deer were present in 1892, when the park is recorded as having an area of about 300 acres (121 hectares), of which only a portion was accessible to the deer (Whittaker, 1892, pp. 182–3). As Whittaker remarked, the common boundary with the Forest of Knaresborough could have been exploited by means of a deer leap to enable wild deer to become trapped in the park – and this underlines the peculiarity of the lack of a license to empark. Red deer joined the fallow deer herd in the park in 1998–99.

FIGURE 46.
The deer park at Ripley from the air. Across the River Nidd to the south (right) a fragment of the co-axial field system is clearly seen and the old meander loop in the floodplain towards the top of the photograph is cut by the embankment of the former railway.

FIGURE 47.
Ripley has one of the few medieval deerparks that still contain deer. The oldest pollard oaks pre-date the park.

In addition to its agricultural uses, the park will have yielded timber, while a little complex of medieval quarries exploited a coarse, purplish Carboniferous sandstone which outcropped to the west of the manor and to the north of the old track to Whipley, around the edge of the wood. Here the thickness of the beds and the angle of dip were ideal, for they obviated the need to extract stone from an ever-deepening pit. Instead, blocks of a suitable size could be split from the bedrock using wedges, slid downs for stacking and sorting in an adjacent administrative area, and then sledged-off for use in the locality. Both churches and the castle appear to have made use of this convenient, if unexceptional supply, while ignoring the more prestigious Magnesian Limestone nearby. The field name Kiln Close, recorded in 1807 in the clayey land later occupied by the northernmost lake implies a brick and tile making industry in the park. In 1733 an agreement was reached with John Davis of Hunslet about making bricks in the 'Horse Pasture', which could be the same place. Some 150,000 'stock' bricks were required, as well as 'water bricks', perhaps for use in drainage works (Ingilby MS, Additional Accession 2662). A much more recent use for the park becomes apparent when one attempts to map the remains of ridge and furrow ploughland there. Suddenly and for no apparent reason a group of ridges will vanish. The answer is found in the economy of the Second World War, when parkland of various kinds was brought under the plough for the first time in centuries because of the food shortages caused by submarine attacks upon convoys.

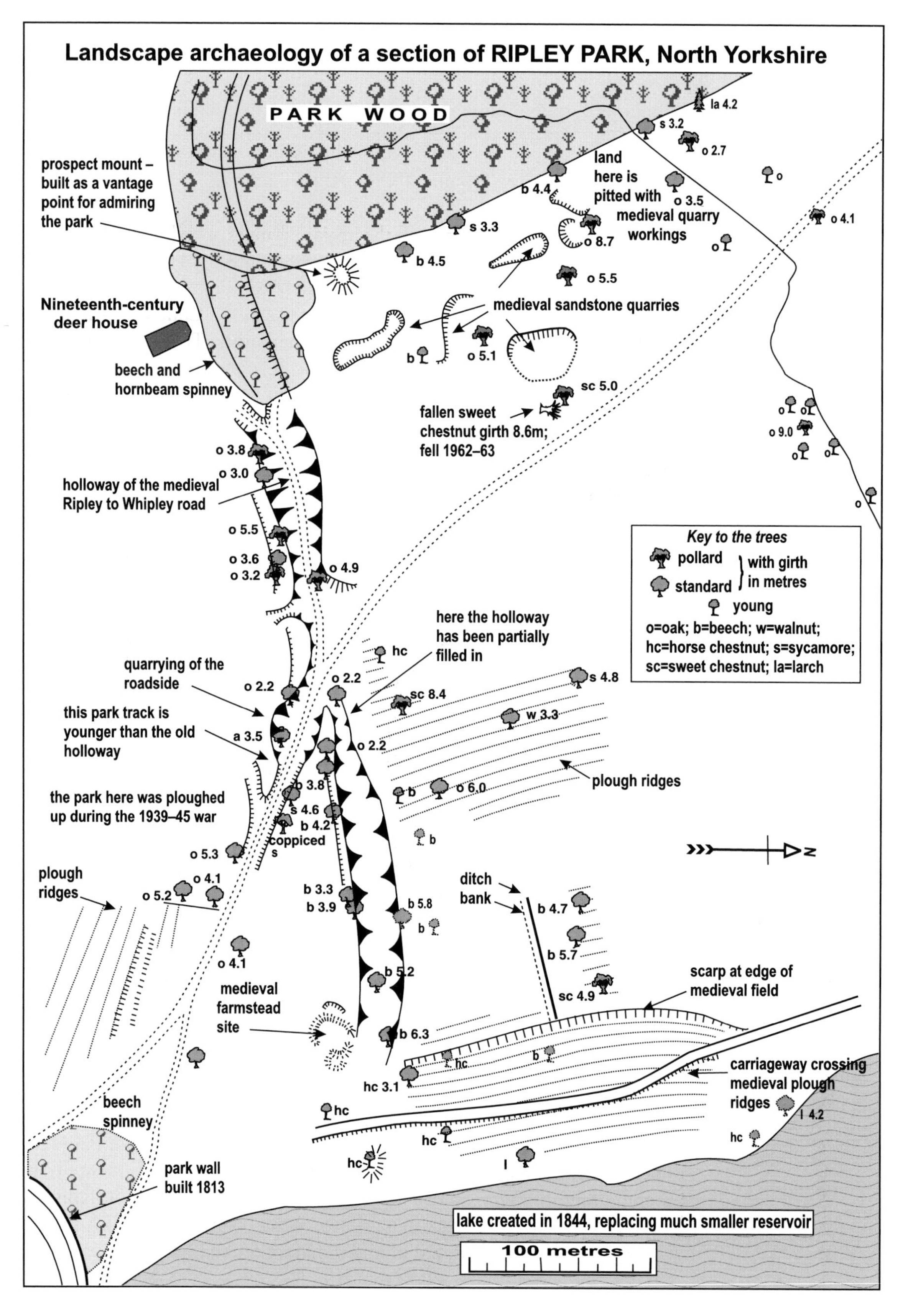
Landscape archaeology of a section of RIPLEY PARK, North Yorkshire
PARK WOOD
prospect mount – built as a vantage point for admiring the park
Nineteenth-century deer house
beech and hornbeam spinney
land here is pitted with medieval quarry workings
medieval sandstone quarries
fallen sweet chestnut girth 8.6m; fell 1962–63
holloway of the medieval Ripley to Whipley road
here the holloway has been partially filled in
quarrying of the roadside
this park track is younger than the old holloway
the park here was ploughed up during the 1939–45 war
coppiced s
plough ridges
plough ridges
ditch
bank
scarp at edge of medieval field
medieval farmstead site
carriageway crossing medieval plough ridges
beech spinney
park wall built 1813
lake created in 1844, replacing much smaller reservoir
100 metres
Key to the trees
pollard
standard
with girth in metres
young
o=oak; b=beech; w=walnut; hc=horse chestnut; s=sycamore; sc=sweet chestnut; la=larch
N

Repercussions

The transformations enacted by the early Ingilbys were not simply confined to the creation of a village and park and the shifting of the church. Each innovation reverberated around the township, causing additional changes and adjustments to follow. Some of the major changes noted must have been inter-linked in some way. When the church location was shifted it was essential that the new site should be accessible not only to the congregation living in Ripley township, but also those of the townships of Clint and Killinghall which combined with Ripley to make up the parish. The new road which sliced across the long established furlong and selion patterns to the south of the village and forked to Killinghall bridge and Nidd provided parishioners in Killinghall with the best possible access to the new church, while at the same time thoroughly disrupted the farming patterns. The curving strip ends that were severed by the road just to the south of the village are still recognisable as earthworks running up to and under the village; apparently, this land was never ploughed again. To the south-west of the road a complete reorganisation of the ridge and furrow patterns was undertaken, with the grain of the new corrugations being made to follow the plan of the road.

The new village must have been populated and it is unlikely that this population came from beyond the township. This survey has revealed the remarkable amount of deserted settlement existing in Ripley, but one cannot tell how much of this represents households who readily or forcibly migrated to the new village, and how much may represent the ravages of the Pestilence or other causes. In the case of the old settlement of Owlcotes, most of the pottery obtained is of the twelfth and thirteenth centuries, with a tailing-off of finds for later medieval centuries. Owlcotes might well have been sacrificed to populate the new village that was closer to the centre of affairs and less entangled in the monastic empire. Alternatively, it might have been weakened by the Black Death and then extinguished in the creation of the deer park, within which it came to lie. Whatever the details of the story may have been, a substantial proportion of the population of the township must have been shifted to people the new settlement.

Although the manor house, which I believe stood on the site of the later castle, was outside the area of open field ploughland, the adjacent village and church were built upon ridge and furrow. On the eastern side of modern Ripley strongly developed broad ridge and furrow can be seen running up to and 'beneath' the settlement. A reorganisation must have been necessary to adjust and compensate for the loss of a substantial amount of ploughland.

To the north and east of the village the changes are less clear-cut. The road from Ripon to Nidderdale may have been diverted to dip into the new village. Certainly it overruns a furlong just to the west of the crossing on Newton Beck, with the direction of ploughing then being turned through 90 degrees in order to make the best use of the severed strip ends. However it is not clear exactly when this took place. The lane from Cayton Grange and

FIGURE 48. Landscape archaeology of a section of Ripley Park, North Yorkshire. The figures by the trees relate to their girths in metres (note the gigantic fallen sweet chestnut, with a girth of 8.6 m)

Birthwaite was diverted sharply to the south-west from its roughly north–south course to cut right across a furlong before straightening and heading directly into the northern end of Ripley. Apparently this was done because the creation of the new village provided a new and more important destination.

These and many lesser responses to the changes were made as the life of the township came to terms with the fundamental changes associated with market, church and village. They are amongst the most difficult details of the landscape to recognise and interpret, giving landscape detection its most intellectually stimulating aspects. They are also vital reminders that everything in the landscape is inter-related, with changes in one component reverberating around the different facets of the cultural landscape until some new form of equilibrium is established.

Enclosure

In 1763–64, Edward Clough, who appears to have had access to more Ingilby papers than survive today, wrote that the decisive moment in the enclosure of Ripley's open fields came in 1494

> ... when Sir William Ingilby ... purchased all the right ?tithe ?title of the Abbot of Convent of Fountains to any Lands Tenements which they had within the Town and Parish of Ripley. These lands were scattered up and down the common open fields, and wd greatly retard the enclosure: unless contrary to their usual maxims, the monks preferred the publick, to their own private advantage: this, considering the time this business took up to complete, does not seem to have been the case.' (Ingilby MS 3757, p. 43).

Such a purchase, freeing the holding of a major landholder, might greatly have facilitated the creation of the deer park. The transfer of lands from John Darnton, Abbot of Fountains is recorded in the Ingilby MSS 228, 229 and 230. Clough, with a characteristically eighteenth-century disdain for rural traditions also wrote that '... towards the close of Henry 8 reign, viz. About 1543 the enclosing of Ripley common fields after struggling with variety of prejudices and jarring interests for more than 100 years together was brought towards a conclusion.' (Ingilby MS 3757, pp. 42–3).

The form of enclosure that was being very actively explored in many parts of England during the closing phases of the medieval period was unlike the Parliamentary Enclosure of the eighteenth and nineteenth centuries in many ways. It was pursued by farmers, great and small, who sought to disentangle their holdings from the intricate interdigitation and scattering of selions associated with open field farming and, by purchase or exchange, to create consolidated holdings. To do so permitted a greater economy of effort, with less time wasted in dragging a plough or harrow between dispersed holdings, and it gave a greater freedom of choice to the farmer who was no longer locked into a common system of crop rotation. But enclosure could also deprive the encloser of the advantages of communal farming and bring disputes about common or private rights of grazing the stubble left after the harvest.

It is clear that a wave of enclosure was sweeping across Nidderdale in the closing phases of the medieval period, and that Clough was correct in suggesting that the changes were 'brought towards a conclusion' by the close of the reign of Henry VIII.

According to Clough (p. 20) the trend began in Ripley at the start of the fifteenth century, with the enclosure of Hallsteads, at the core of the ploughland of the manor, by 1413 (this being the movement, mentioned above, that might have been associated with the foundation of the village). Enclosed land of a kind existed long before this, for some of the assarted land was cleared by individuals and stood apart from the common fields when they worked it, though there was sometimes confusion about whether it remained part of the common waste in the season after the crops had been taken in. Around 1250–78 William de Ripley gained the right to common grazing in 'wasta mora' whenever Robert Stopham had taken his crop away, while Robert gained the right to enclose land below Newton that had belonged to William. Since the land was called Edmundriding and had thus apparently been assarted by an Edmund it is not clear why it was not already enclosed. Equally, it is not clear how William could have an exclusive right to grazing that was also common (Ingilby MS 178).

Dozens of medieval land exchanges produced the field patterns associated with the countryside of Ripley as it is seen today. Very few of these are recorded, but one of the last occurred in 1538, when Sir William Ingilby and Robert Ripley of Birthwaite exchanged strips in the field of Ripley and Sir William agreed to fence the land from the highway on the western side between: 'Robert's land now enclosed in the east field of Ripley and the lands next adjoining now in the tenure of Thomas Lang to the bottom of Elom ...' (Ingilby MS 1066). The fences, hedges and walls used to mark the property boundaries between the parcels of selions that had been gathered together by swops and sales followed the sweeping edges of reversed-'S' or reversed-'C' strip plans. Not only did they produce more pleasing curves in the countryside vistas, they also helped to perpetuate the medieval ploughland patterns, often surviving after modern ploughing has levelled the tell-tale corrugations of ridge and furrow. In reconstructing the medieval landscape of the township (Plate 7) such field boundary evidence is at least as valuable as the information derived from ridge and furrow.

In the sixteenth century, population was recovering strongly from the onslaughts of the Pestilence, while the suppression of the monasteries released land for investors and speculators, who might be locally based or outsiders. The Ingilby family acquired extensive estates further up the dale at Dacre and Hartwith, while within the township of Ripley late fifteenth-century arrangements and an agreement between Sir William Ingilby and the Abbot of Fountains arbitrated by a George Strangeways, bachelor of sacred theology, and Ranulph Pigot for an exchange of territory (Ingilby MS 231) had left the family in control of most of the farmland in the township. 'Only seventy-one acres of land in the township were freehold. The remaining 1,000 or so acres of enclosed land, except for fifty-six acres of glebe, belonged to the lord of the manor. The park, and the desmesne farmed directly by the Ingilbys, accounted for over 500 acres, and the rest was let to tenants on leases of twenty-one years, (Jennings, p. 128).

By the reigns of the Stuart kings in the seventeenth century, the hedges resulting from the Tudor enclosures were well-established. The hedgerows, like the barbed wire on the ranges of the American West, had compartmentalised the open countrysides and made it harder to travel freely. Even so one would not expect political explanations for the hedge-breaking that seems to have been endemic in Ripley. Some of the offences seem bizarre and perverse: 'Robert Hodgson broke the hedge of a certain close called Chappell Flatt and placed the horse of Thomas Withes therein' (Ingilby MS 1607, no. 4, 10 April 1627 [3 Ch. I]). Most are unexplained: 'Katherine Scarelett, widow, broke the hedges of Mr William Pulleyn, clerk, [the rector] in a close called *Parsonflatt* at Birthwait' (Ingilby MS 1607, no. 28, 15 April 21 Js. I). In some cases, the motivation seems to have been the stealing of fuel

> Thomas Holdsworth cut down a tree called in English a Crabtree in Robert Bransbye's hedge (Ingilby MS 1607, no. 13, 16 April 1649).
>
> William Knowles cut down and carried off an elder in *le Chappleflatt* (Ingilby MS 1608, no. 5, 17 April 1629).
>
> Thomas Kidd allowed his children to cut down and carry off wood, in English *twistinge of oake and ashe*, from Ripley Park (*ibid*).

The sorry nature of these little offences makes one wonder if there might be some association with a Little Ice Age around the Elizabethan/Stuart transition which could have intensified the need for fuel? The cases cited were heard in April, and earlier in spring the legal reserves of fuel could have been exhausted and a cold snap might have driven some people to steal from park and hedgerow.

Other cases heard by the seventeenth-century manor court concerned overgrown rather than depleted hedgerows: 'A paine laid that John Browne and Francis Launsdell shall cutt a peece of hedge alonge the street [Ms torn] Ripon that men may passe with corne and hay over against Langbrookes before Candelmas next' (Ingilby MS 1608, no. 2, 9 October 1627).

Paralleling the movement to enclose and thus privatise the land in Tudor times was a shift from a form of farming centred on ploughing to one based on grazing. It was in the closing stages of the medieval period that the characteristic Nidderdale countrysides were formed. The fieldscape is composed of enclosed open field selions and meadow doles, assarts and intakes, some walled, but those in the gentler countrysides being hemmed by curving hedgerows studded with old oak and ash pollards and standards. The situation in Wensleydale, which shares some similarities were explained by William Marshall in 1796:

> The enclosed parts of this neighbourhood, when seen at some distance, have the appearance of woodlands; the inclosures being mostly narrow and full of hedgerow timber. The age, on par, is about fifty years. In half a century more, the value of timber of some parts of it, if suffered to stand will probably be equal to the value of the land ... In this country it seems to be a general idea, founded perhaps on experience, that lofty hedgerows are beneficial to grassland: increasing its productivness by their warmth and giving shelter and shade to pasturing stock ...

He alluded particularly to the planting of trees in hedges surrounding newly-enclosed land and thought the advantages to be:

> ... far superior to any disadvantage accruing therefrom, even when they have been suffered to grow in a state of almost total neglect. Land which has lain open, and which has been kept in a state of aeration during a succession of ages, is equally productive of grass and trees. And it is generally good management to let it lie to grass for some length of time after inclosure'.

He thought that the trees flourished with unusual vigour in the newly-inclosed land of arable fields, but that the ash should be eliminated from hedgerows and replaced with oak beside fields subject to arable cultivation because its roots obstructed the plough.

As the wooded waste and the wood pastures receded in the phase of assarting during the twelfth and thirteenth centuries, the timber deficit must have been countered by growing the oak, ash and elm as pollards in the hedgerows surrounding the assarts. Then, as the old common meadows and ploughlands were enclosed in the fourteenth and fifteenth centuries, similar tree-studded hedgerows were planted around the new, privately tenanted fields. Consequently, in most distant views, Nidderdale must have resembled a great wildwood. From the first editions of the Ordnance Survey six inch to one mile maps, which date from the middle of the nineteenth century, it is plain that then this wooded aspect was much more pronounced than today. Pollards in the dale were recently studied, the best concentrations being in the Ripley-Nidd-Clint-Birstwith region, and it was seen that 68 per cent of 186 ancient trees stand in hedgerows that still survive, and a further 32 per cent stood in hedgerows that have been destroyed (Muir, 2000 p. 100). While recognising the numerous problems associated with producing ages for hollow veteran trees, the evidence from Nidderdale does seem to support Marshall's notion of a surge of planting in the earlier part of the eighteenth century, though a great many trees and hedgerows derive from earlier phases of planting, going back eight centuries or more. The popular medieval practice of pollarding hedgerow trees to produce periodic crops of poles and fodder seems to have continued until the middle of the eighteenth century. In 1759 the route from Knaresborough to Pateley Bridge was turnpiked, with a widening and straightening of its course. No pollards are found in its hedgerows except just to the west of Scarah Bridge, where the narrower original road seems to have been retained.

Were a villager of modern Ripley to be transported back through time and released into the township as it was at the time of Domesday Book they would feel hopelessly lost, threatened and disorientated. On the other hand, were this villager to step into Ripley at the time of Henry VIII, familiar sights and passages of scenery would nurture a sense of belonging over one of strangeness. The church would be recognisable, and so would the village street plan and the main divisions of the land. The countryside would contain more young hedges, the roads would be narrow, stoney tracks and outside the village, most of the dwellings of timber and thatch would lack an upper storey . Even so, there would be no mistaking directions or the right paths to travel. Though it is often stigmatised as a time of introspection and stagnation the medieval period, and particularly is closing centuries, was a time of change when many of the landmarks of today's countrysides were put in place.

References

Farrer, W., *Early Yorkshire Charters*, vol. I, Edinburgh, 1914.

Sir Thomas Ingilby, *Ripley Castle.*

Jennings, B. (ed.), *A History of Nidderdale*, 2nd edn, Advertiser Press, Huddersfield, 1983.

Lancaster, W. T., *The Early History of Ripley and the Ingilby Family*, John Whitehead, Leeds, 1918.

McCutcheon, K. L., *Yorkshire Fairs and Markets* (Thoresby Soc., 39).

Muir, R., 'Village Evolution in the Yorkshire Dales', *Northern History*, 35, 1998, pp. 1–16.

Neave, S., *Medieval Parks of East Yorkshire*, Hull, 1991.

Rackham, O., *History of the Countryside*, Weidenfeld & Nicholson, London, 1994.

Thorpe, J., *Ripley: Its History and Antiquities*, Whitaker, London 1866.

Whitaker, J., *Deer Parks and Paddocks of England*, Ballantyne, Hanson & Co., London, 1892.

CHAPTER FIVE

A Time for Kneeling

With the passing of the Middle Ages, the impress of lordly control upon the landscape was not diminished, indeed, it became stronger. It was as though the landscapes were kneeling before their masters and changing at their will. The world of the feudal tenant had already given way to that of the copyholder and the harshest of the medieval injustices had been removed. Even so, one might argue that during the medieval period, peasant toil was the great engine that shaped the countryside. Occasionally, lords or their agents might intervene to consolidate their demesnes or insert a deer park into the agricultural scene, but on the whole it was the ordinary people working through the institutions of the manor who decided how the scenery took its shape. During the post-medieval centuries, however, in our locality and many others, the local aristocracy were the main initiators of change. They were guided by vogues in landscaping and gardening, by the fashionable quest for 'Improvements' in agriculture with would produce greater financial efficiency and, sometimes, by concerns to improve conditions for their poorer tenants. Often these goals were in conflict, but the aristocrats and greater squires of the sixteenth, seventeenth, eighteeenth, and nineteenth centuries asserted their presence in the countryside with rare confidence and mastery. They owned the land all around the scenes of their innovations and transformations and there were no planning officers or planning committees to constrain their enthusiasms and temper their eccentricities. Surprisingly – or perhaps not surprisingly – the buildings and landscapes that they created seem to have embodied more taste and élan than modern equivalents that have enjoyed the attentions of the appointed planning cogniscenti.

The castle

By even the most lengthy of the different dating formulations, the medieval period was over before the castle at Ripley was begun. The age of the castle was over too, and had been for some time. A large part of the English aristocracy had perished during the Wars of the Roses, and when the new Tudor dynasty emerged victorious, England entered upon an era of strong, centralised government. Since no rival could hope successfully to defy royal authority and the invincible mercenary armies supporting it, the private castle became redundant as a stronghold. It still had a very powerful symbolic

FIGURE 49.
In Nidderdale, as elsewhere, ordinary people tended to take their lead from their lords. As royalists and devout Catholics the Ingilbys encountered dangerous times in the 1640s. Bullet holes in the east wall of the church are said to result from the actions of Cromwell's firing squads.

association with aristocratic status, so that when members of the nobility abandoned their castles and turned towards homes built to provide comfort and accommodation for great gatherings of guests, aspects borrowed from the old castles were sometimes incorporated.

Ripley Castle is well documented, and so only a brief description is provided here. In 1548 Sir William Ingilby, who had campaigned in Scotland with Edwin Seymour, the Earl of Hertford, began work on a fortified tower at Ripley. The extent to which it was a fortification rather than a symbol of status must be queried, for in 1567 he raised the height of the church tower, nearby. In so doing, he improved the capabilities of a vantage point and firing platform which might have been used to sweep his secular tower ramparts of defenders had an adversary ever come to Ripley. If the physical rather than the psychological trappings of defence were being sought, then the works might have kept at bay a small Scottish raiding party or a more local band of insurgents – but little more. The work may be put in context if we look at some of the other castles that were being built in England at this time – though on rather grander scales.

Despite the excitement, even confidence, of England during the Renaissance, the kingdom was littered with ruins, some of them monastic, but many of them those of abandoned castles. Thompson made a survey of castles mentioned in Leland's itinery of 1535–45. In 258 cases some indication of the condition of a castle was provided, and of these just 91 were in normal use, 30 were partially derelict, while 137 were ruinous (Thompson, 1987, p. 104). In the time of Henry VIII the old strategic circumstance of a confrontation between royal and baronial castles was replaced by one in which the castle, fort or blockhouse was an instrument of national defence, with the strongholds being employed to armour the invasion coast of the English Channel. A systematic programme of anti-invasion defences in the form of a coastal chain of artillery forts was begun in 1539. Even castles built a good century before Ripley's tower had few serious defensive pretensions. Hurstmonceux in Sussex of 1440 was a spectacular building set within a great moat, yet its walls were thin, of brick and breached by great mullioned windows, and although the cross cut arrow slits in its gatehouse turrets were symbols of war the castle could scarcely have resisted siege engines for long.

About the time that Hurstmonceux was built, the manorial focus at Ripley will have been a fortified manor house with a hall, in the traditional style, that was protected by a gatehouse of stone that still survives. The gatehouse was built by Sir John Ingilby, 1434–57 and could have been a genuinely defensive response to the lawlessness and insecurity of the reign of Henry VI – though in a status-conscious age it also made a statement about the status of the owner. The domestic element has gone, demolished during a rebuilding in the late eighteenth century, but from Pennant, who saw it in 1773, just a few years before it perished, we have a valuable description. He noted that the house that he saw was partly a tower of the sixteenth century, but that

> ... a more ancient edifice still remains of wood and plaister, and solid wooden stairs ... The entrance to the house is through a porch, the descent into it by three steps; the hall is large and lofty, has its bow

FIGURE 50. The late medieval gatehouse at Ripley Castle.

FIGURE 51.
Ripley Castle, with the tower of 1555 on the right. The large window openings show that comfort and aesthetic aspects exceeded concerns with defence.

windows, its elevated upper table and its table for vassals, and is floored with brick' (Pennant, 1804, pp. 115–16).

From this it is clear that an essentially medieval hall, open to the rafters high above and with its dais and high table for the lords and their guests overlooking the tables used by servants and tenants, survived at Ripley till the end of the 1780s. Had it survived to the present day it would be one of very few unchanged examples and would be a magnet for visitors and scholars. There is no mention and no indication of a moat, though some other manors in the district, like the earlier Markenfield and the later Clint, were moated – as was the much earlier building at Sadler Carr.

Sir William's tower was added to the fortified medieval manor house. It was a building that echoed the archaic defensive architecture of the Norman donjon or keep – a simple form that enjoyed a flurry of popularity in the dying days of the medieval castle. As with the older gatehouse, it had no credibility in terms of contemporary warfare, but could offer protection in times of raiding or local unrest. The buttressed and crenellated tower abutted against the manor house but was entirely self-contained and had three floors. It took seven years to build, and when it was finished its proud owner installed a wooden inscription in the Knight's Chamber on the third floor: 'In the year of our Lord 1555 was this house builded by me, Sir William Ingilby, Knight; Philip and Mary reigning that time'.

The home that might have been considered impressive in Tudor and Elizabethan times was, by the eighteenth century, distinctly unfashionable. Sir John Ingilby (1757–1815) married Elizabeth Amcotts in 1780 and £12,000 was borrowed to fund the rebuilding of Ripley Castle; before the work began, Sir John commissioned paintings showing four vistas of the castle and village. Here one can see the stone keep and manor house with the mullioned windows set beside a lawned courtyard patrolled by peacocks. Although Sir John intended to preserve as much of the old structures as possible, it was found that the buildings were in a worse condition than anticipated. Sir John wrote to Somerset Herald:

> ... when we began upon my old mansion it was in such wretched condition that we were obliged to do a great deal more to it than was first intended – I was determined upon preserving as much as possible of the old place and by that means have spoiled my plan in the opinion of some people, but not withstanding the inconveniences of our Ancestor's buildings I prefer them to the modern structures. Any man who has money can build a house, but few can shew the same house his family has lived in so many years as the Ingilbys have done at Ripley (Ingilby 1784)

Demolition work on the old manor house in 1784 cost £10, while the rebuilding was completed in 1786 at a cost of £678 13*s.* 4*d.* (£678.66), with crenellations being added to the side ranges of the gatehouse a couple of years later. With the work almost complete, however, Sir Wharton Amcotts refused to part with his daughter's dowry and in 1794 the couple and their children, having greatly overspent, were obliged to leave the castle and seek refuge on the continent. Elizabeth abandoned the family and went to live with her aunt in Lincolnshire in 1801, though in 1804, drawing upon the valuable timber reserves of the estate, Sir John paid off the last of his creditors and returned to Ripley. Building soon began anew, and in 1807–12 three blocks of stables and outbuildings were erected.

Sir John's salvation must have owed much to the seemingly manic energy that he had devoted to the afforestation of the estate, though any mature timber sold must have been planted by his predecessors. Fortunately, he kept a record of the trees that he planted from 1781 (Ingilby MS 2838). Some were garden trees, some formed shrubberies, but some were planted in very large numbers, thus in 1781

> Planted upon Tom Hill (upon Scara Moor) oaks, ash, Beech, birch, Holley a great quantity and 2000 Scotch firs 700 Spruce and 600 Larches and 1000 Sycamores' Tom Hill had not seen the end of planting, for in 1782: 'Planted upon Tom Hill on Scara Moor 300 Beech 900 Dutch Elm 1000 Scotch Firs 500 Spruce ... 500 Larches 100 Hornbeams 200 Lymes 200 wild black cherries ...' Plantations were established in all suitable corners of the estate, and also in places less accommodating: 'March 15 1783 Planted 140 large oaks in broxholm as before [all died]'.

The records show that in November and December 1783 Sir John planted some 6,538 trees of a dozen different species (Ingilby MS 2839).

Faced with adversity, he was able to harvest very considerable resources of timber, much of it coming from compartments of the park like High Rails, Lodge Wood and Hollin Bank (Holly Bank). In 1800 he wrote:

> You will be satisfied I was in the right about the Wood and I am resolved to cut down as much [trees] as possible, without very much defacing Ripley, that I may get my affirs as clear as I can and as firm as I can – Let High Rails and the Lodge Wood be both sold, these are already fenced off with a little care the moor will ?thrive/live again, and High rails perhaps must be new planted ?some surface drainage – a very trifle ... I think if it was plowed and sown with mast, acorns ... it wd come quicker and better however the walk from gate to gate must be left, there are a great many trees in Hollin bank from the lane gate to the End of ?Deanloves Pasture that may come down – I leave all this to your judgement ...' (Ingilby MS, additional letter 37 Accession 2662).

Extensive sales of timber allowed Sir John to return and the initiative for improvement was regained. During the nineteenth century the castle was generally regarded as a building of some quality in a setting that many might envy

> During the early part of the present century the Castle was much enlarged. It is now a spacious and commodious mansion indicative of unsuspecting security, and having the adjuncts of modern greenhouses and hothouses rather than of ancient stockades and dungeons deep; the greenhouses and hothouses being scarcely excelled in the kingdom. The "Castle" is a mere effort of whim, said to have been embattled "for ornament," but as a matter of fact, rendered more ridiculous than ornamental. The lodge and the great tower still, however, retain their characteristics of strength and security, and so far keep alive the old traditions of local tumults and foreign dangers. The site of the Castle is capable of great picturesqueness, which has not been entirely neglected (Wheater, 1888 p. 211).

The castle and its gardens, like most aristocratic preserves elsewhere in England, were funded by the rents of tenants on the estate(s). Their manner of living was quite different and the homes that they occupied in the century following the close of the medieval period were very different from the seemingly timeless cottages and farmsteads seen by modern visitors to Nidderdale. The older stone buildings still standing in the countryside are an inheritance from the eighteenth and nineteenth centuries, though the countryfolk of the sixteenth and seventeenth centuries occupied dwellings of timber, lath and plaster and thatch. These were generally single-storey houses built according to the techniques of the medieval cruck-framing technique. The essential structure of the house was provided by 'A'-shaped frames created by splitting curving oak boughs to form two similar blades and then joining the blades near their tips. Each 'A'-frame defined the end of a bay, so that a house of four crucks or four pairs of posts had three bays. (The Ripley records refer to pairs of crucks or crocks and pairs of posts. It is not clear if 'pairs of posts' refers to box-framing rather than cruck-framing. It probably does, for in the entry for John Nurssa the word 'posts' was deleted and replaced by 'crockes'.) Roofing was normally of thatch and the crucks were linked by wall plates, purlins and tie beams to form a structure, the gaps in which could be filled by studs and plaster panels. Later some of these houses were encased in stone and some such examples still survive.

Details of housing in the north-western, Scarah district of Ripley can be found in a survey produced for the landlord in 1635 (Ingilby MS 2453). Some tenants had tiny holdings and must have depended heavily on the nearby common, and even some of the largest of the tenancies would be uneconomic by the standards of today. The entry for one of the larger tenants reads as follows:

> Anthony Raynard: House of 7 pairs of posts, 2 barns and 2 little houses in good repair, also a laith [barn] in the field, 3 chambers and a parlour and a wain [cart] house £7 10*s.* 0*d.* [£7.50]. He holds 25a-0r-24p of land [about 10 ha.] worth £15 12*s.* 0*d.* [£15.60 per annum]

Raynard's house of seven pairs of posts or six bays was a long one. The width of a cruck-framed house was determined by the length of the curving boughs that could be obtained. Houses could not be widened and so they were enlarged lengthwise by adding additional bays.

The reference to William Raynardson's house still employs the medieval hall concept to describe the main room, while like a high proportion of other Ripley people, he kept a small apple orchard

> William Raynardson: one hall howse, a kitchin, 4 parlours, 4 chambers, one stable, one smithie, a barne and a kilne, all of crockes, in good repaire, 6 aple trees. £01 13*s.* 4*d.* [about £1.16p]

Christopher Lofthouse had a more modest holding:

> Christopher Loftus: Dwelling house of 4 pairs of posts, 2 parlours, 2 chambers, one barn in good repair, one oxhouse wanteth thack [thatch]£4. He holds 18a-0r-17p of land [about 7 ha.] worth £8 p.a.

Others were worse off:

> Widow Tompson: House of 4 pairs of posts, 2 parlours, 2 chambers, 1 laith, decayed in walls and thack, one oven house and laith in the field in repair £6 10*s.* 0*d.* [£6.50]

Widow Thompson may well have been the widow of one William Thompson of Scarah Bank, who died in 1621. If so, the tenanted land must have been relinquished by his widow. His probate inventory (a legal listing of all that he possessed at the time of his death) is as follows:

8 oxen £24; 2 young stotts [bull calfs] £2 3*s.* 8*d.*; 2 young stirks [cattle of both sexes aged 1–2 years] £1 6*s.* 8*d.*; 8 kine [cows] £18 13*s.* 4*d.*; 4 heifers £6 13*s.* 4*d.*; 3 stirks £1 10*s.* 0*d.*; 3 young calves £1; 2 mares £4; 27 sheep £6; Hay £3; Hay at Beckside £4 13*s.* 4*d.*; rye, beans, barley, malt, wheat £7; sown corn £1; 1 wain [cart] 30/– [£1.50]; 1 coup 10/– [50p]; plough gear, harrow, 1 ox, 2 old sleets [? The dialect word normally refers to a flat field. Sleds?] £2 3*s.* 0*d.*; Tithe of Farm at Scarah £40; wool, 1 wooden web 8/–; 1 linen web at the websters 12/– (*Township of Clint-cum-Hamlets*)

It is plain from more detailed inventories of the late seventeenth century that tenants were able to survive on holdings of just a few acres because they combined cottage textile industries, both linen and wool, with their farming and cheese-making. It is also clear that the modern habit of designating rooms for different specialised functions did not exist in the seventeenth century. Dairying equipment, beds, looms and spinning gear were apparently jumbled together, while householders would commonly receive guests from their beds. In these inventories, the main room of the farmstead is frequently referred to as the 'house', a legacy of the all-purpose hall of the medieval house. Other rooms were commonly termed 'parlour' and 'chamber', though food such as pies and cheeses might be kept in the chamber along with the beds and the parlour could have a dining table, bedstead, newly-woven cloth and a variety of tubs and cupboards. There could be a more specialised room in the 'low

FIGURE 52.
The view across the former garden towards the castle. The great retaining bank of the main pond runs diagonally across the central area of the photograph from bottom right to top left.

end' of the house with churns, presses, bowls and other items of cheese-making equipment as well as brewing gear. The next generation of farmsteads were the laithe-houses: stone-built houses of two storeys with an adjoining barn combined under the same roof. Unlike the preceding generations of cottages they were built to last and at some expense. Many survive to this day.

The Parson's garden

The programme of study recorded in this book began much less ambitiously as a survey of ancient trees and earthworks in the Castle park. While carrying equipment to and from an entrance to the park I became aware of a substantial embankment in the pasture to the south of the churchyard. Having gained permission to enter the pasture it was immediately apparent that it was patterned with the earthworks of a former garden. When the earthworks were surveyed it was equally plain that the garden was one of the formal type and that stacked ponds had formed its most prominent features. However, no relevant documentation could be found, so that the lay-out of the earthworks provided the main evidence.

In cases such as this, dating evidence may come from two sources. Documents may not mention a feature, but the very lack of a mention may signify that the feature concerned did not then exist. If the evidence for non-existence can be used to provide temporal brackets, then it can be argued that the feature *did* exist at some time within those brackets. Secondly, stylistic evidence can be invoked, for fashions in garden design have changed significantly over the centuries. Here one must be careful, for with gardens, as with architecture, designs could become outmoded in the fashionable south-east of England, yet still be favoured in the north.

In 1523 the area concerned was not a garden, but the parson's orchard, for in this year Sir William Ingilby leased to parson Thomas Skawesby a close beside the mill beck which was said to lie between the park and Skawesby's orchard. Evidence for 1597 hints that a water garden might have been created, for the manor court instructed William Pulleyn, the rector to clean his ditch or water sewer at *Le Stanke*, as he had been ordered to do by a previous pain (Ingilby MS 1607, no. 2, 20 October 39 Eliz.) The words '*le Stanke*' refer to a pond; its location is uncertain, although 18 years later (Ingilby MS 1607, no 9, 10 October 1615) the same rector was in the mercy of the court for not scouring his ditch in the churchyard (very close to the garden setting).

However, stronger evidence shows that the garden had still not been created, for an enquiry showed that the garden-to-be was still an orchard: 'The foresaid jurors, concerning the view of a certain parcel of land called Walkemill Inge in this manor, lying at the south part of the orchard called Parsons Orchard, on the east part of the stream there ...' (Ingilby MS 1608, no. 8, 22 April 1631). Orchards still dominated the scene between the churchyard and the beck in 1650, when the manor court instructed the next rector, Leonard Campleton

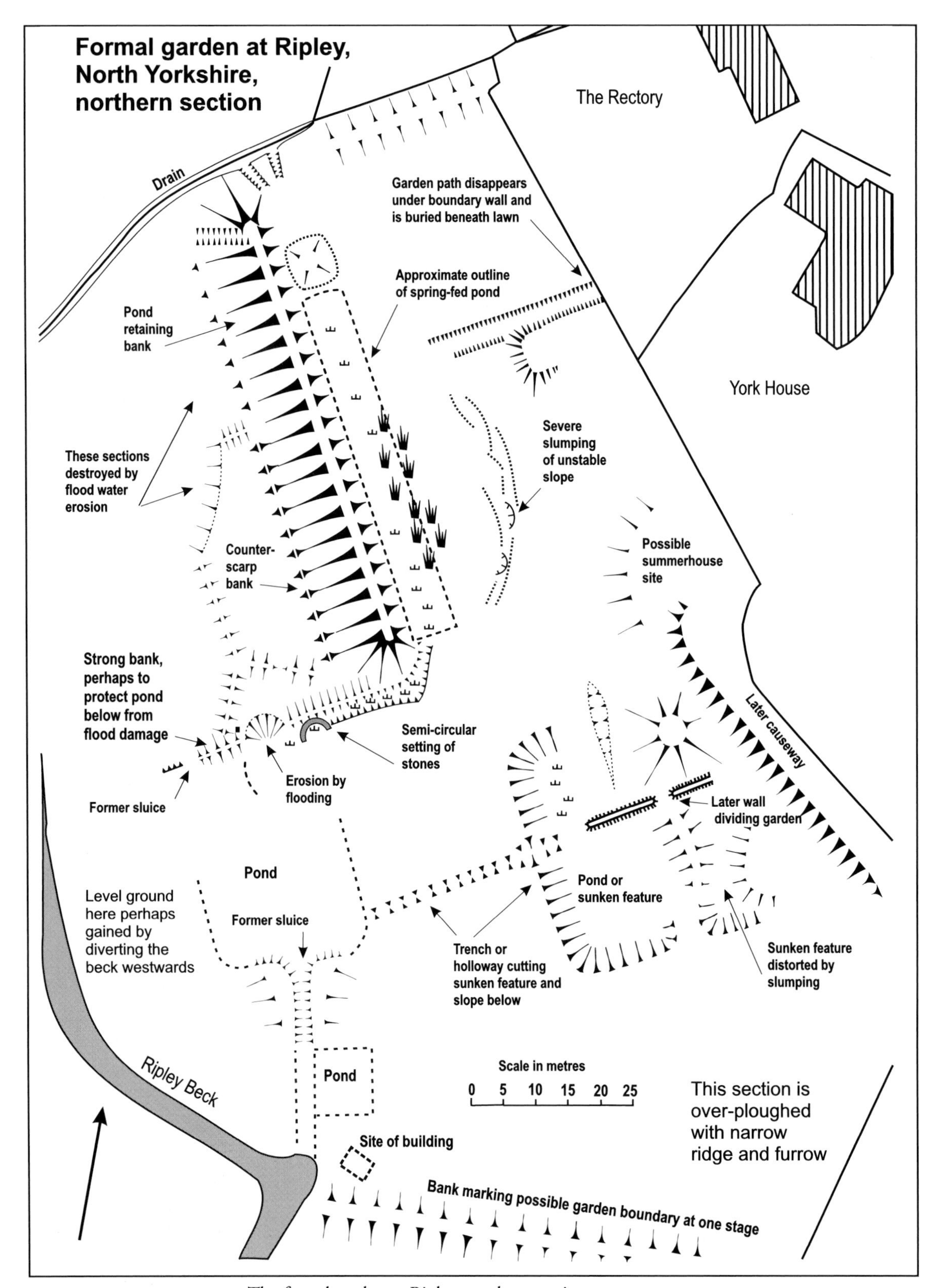

FIGURE 53. The formal garden at Ripley: northern section.

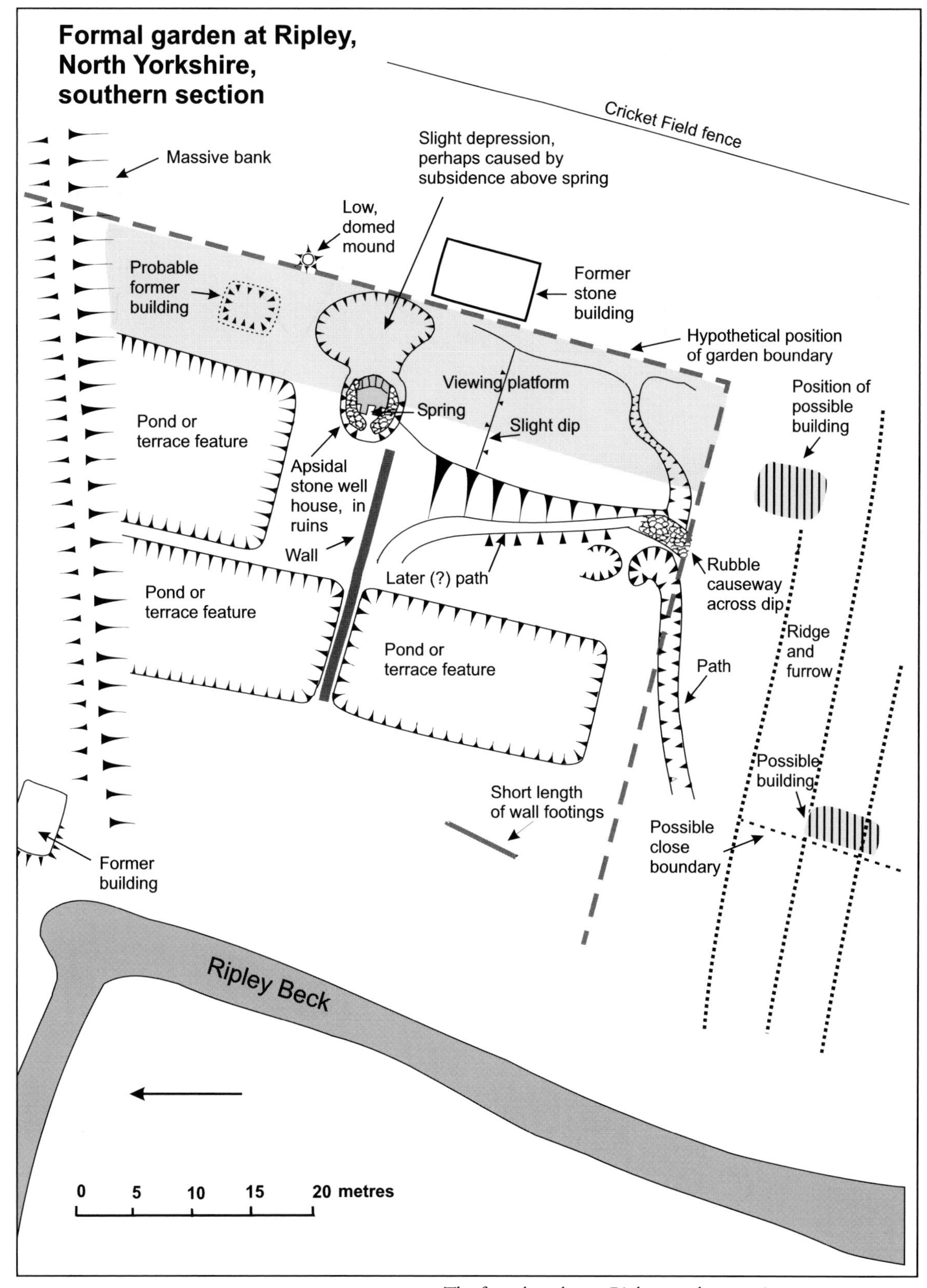

FIGURE 54. The formal garden at Ripley: southern section.

to scour his ditch between the churchyard and his orchards (Ingilby MS 1607, no. 15, 6 May 1650).

In 1752 Chippendale surveyed the estate and the field containing the garden remains was simply captioned 'Parson's Pasture'. The documentation produces time brackets of 1650–1752, which are rather wide, though they can be narrowed by inference. The total absence of any reference to the former existence of a magnificent sweep of gardens on the Chippendale map strongly hints that the gardens were long extinct and covered in grassland by then.

FIGURE 55.
An aerial view of Ripley, with the earthworks of the formal garden in the centre of the photograph, to the right of the village.
© ENGLISH HERITAGE

Turning to the stylistic evidence, the most likely temporal abode for our garden is the seventeenth century. The gardens of this period tended to be geometrical and of relatively unambitious proportions: 'The total garden area was usually modest, between half an acre and 2 acres, though occasionally rising to 4 or 5' (Thacker, 1994, p. 110). The Ripley garden comes in at around the upper limit quoted. Thacker went on to note the words of Sir William Temple in 1685 who considered English and Dutch gardens to be smaller than those of Italy, seldom exceeding four, six or eight acres within their enclosing walls (p. 112). A garden newly created at Hillesden in Buckinghamshire in the 1640s had features similar to our own. The falling ground in front of the house was terraced, with one terrace overlooking a rectangular pool or canal. An earlier garden, built partly on sloping ground at Holdenby in Northamptonshire by Sir Christopher Hatton in 1579–87 also echoed some features.

The sloping topography of the Ripley site was conducive to the use of the flights of terraces so popular in seventeenth-century designs, though the curving nature of the flanks of the Ripley Beck introduced complications. It is conceivable that in creating the garden the beck was diverted to the south-west to create a little more level space at the foot of the slope. The hydrology, with springs seeping from the 'watter sands' and the beck at the foot of the garden, provided excellent opportunities for pond-making. The main pond, elongated with a massive retaining bank, was fed by a spring rising from its floor. Some of the others may have been filled in this manner, one cannot be sure, for such springs migrate. There is also a possibility that some of the sunken or embanked rectangular features that seem to have been ponds were actually sunken lawns or beds. Taking the stylistic evidence into account, a date for the Ripley garden of around 1650 would seem appropriate; much later and the fashions would change.

Its state of preservation varies from place to place, some earthworks seeming quite fresh, as with the retaining bank and its crestline walkway, while elsewhere the instability of the deposits, as on the ground immediately overlooking the great pond, has resulted in extensive slumping. On the floodplain at the foot of the garden, flooding by the Ripley Beck has carried away sections of low retaining banks nearest to the stream. After the garden fell into disuse a stone wall, probably of an agricultural nature was built to bisect two rectangular ponds or sunken features. The south-eastern sector of the garden was overploughed, perhaps in the eighteenth century, leaving faint traces of narrow ridge and furrow.

The land upon which the garden was built must surely have been a part of the fragmented glebe. The nature of the site is such that the slopes form a curve, allowing a viewer in the extreme south-east to enjoy a sweeping panorama of the ponds, beds and lawns. Here, a level area may have served as a viewing platform, while at its margin are the remains of a well or conduit house. Too little is left to allow precise dating, but it resembles conduit houses of the mid-sixteenth century (S. Moorhouse, personal communication). From the surviving stonework it could be argued that the original building was of

dressed stone and apsidal with a corbelled roof. A spring still issues from its base and this water might have been used in the garden, whether or not this was the original function of the structure.

The garden exemplifies the perils of easy and hasty assumptions. On discovering a 'lost' garden within yards of a stately home most investigators would assume that the mansion and the garden must be associated. Were this assumption made, then one mistake could pave the way for another. The garden designer, Peter Aram, was gardener at Newby Hall and then at Ripley Castle in the eighteenth century. In the 1720s he wrote an unpublished work, *A Treatise Concerning the Making and Management of Gardens*, while amongst the Ingilby papers is a treatise on flowers by Aram (Ingilby 3664). His designs were severely formal, and in Yorkshire, away from the fashionable heartlands, gardens of this type were built until the 1730s (Sheeran, 1990, pp. 24–5). How easy, then, it might be to assume that our gardens were attached to Ripley Castle and designed by Peter Aram in a final flourish of the formal style.

The reality was quite different. In the days of the formal garden, viewing was intended to take place from various favoured positions – sometimes a mount was built to serve as a viewing platform – and the modern enjoyment of unfolding vistas as one walks through a park or garden found less favour. Wherever possible, formal gardens were placed to the south of their attendant mansions, while the geometry of the garden was designed to link the garden to the house from which it was viewed. Our garden was poorly-placed to be enjoyed from the castle – but very well placed to be enjoyed from the rectory site. When we project the geometry of the garden northwards into the rectory plot, the lines do not converge on any existing building but they do indicate where a former rectory, the one associated with the garden, had stood. It had gone completely by 1752, further evidence of the early character of the garden.

The garden was the creation of a rector of Ripley church who must have been a man of considerable wealth. Apart from the costs of creating the garden earthworks and stocking the beds with plants, there will have been the high costs of weeding and maintaining this large expanse of manicured countryside. Formal gardens demanded more maintenance than the more relaxed creations of the eighteenth century and there was no recourse to weedkillers or pesticides. Any successor whose pocket was less deep would have found this garden an impossible burden, so its life-span may have been quite short. Were it possible to refine the dating evidence then it would be possible to identify the garden maker. The rectors of Ripley are known and he should be one of the following:

1631	Leonard Campleton
1655	Nathaniel Raithband
1660–84	John Kirshaw

However, one cannot completely rule out the possibility of an Elizabethan garden that was lost early in the seventeenth century – as suggested by the reference to rector Pulleyn's pond in 1597 and the style of the conduit house ruins.

FIGURE 56.
The garden from the air. The earthworks in the centre of the photograph, below the church and castle.

FIGURE 57.
Floodwater partly filling one of the small ponds at the foot of the garden: a former sluice runs directly towards the camera.

FIGURE 58.
The ruins of the well house or conduit house, perhaps of the seventeenth century amongst the earthworks of the garden.

There is one last point of interest. The garden, with some quite pronounced earthworks, lay quite unsuspected within a stone's throw of Ripley Castle, a quite considerable tourist attraction and a magnet, one would suppose, for historians until discovered during this survey in the late 1990s. This fact must dispel any notion that worthwhile discoveries about the England's heritage may only exist in remote and desolate places. A large public car park, complete with toilets, directly overlooks the garden and many of its diagnostic features are visible from here and also from the popular footpath leading to the cascade and Hollybank Lane.

The park

Despite the existence of long-established boundaries that were demarcated by prominent bank and ditch earthworks along the eastern side of Dob Lane and the western edge of Holly Bank Wood, the limits of the park fluctuated considerably during the centuries following its late medieval creation. In the seventeenth century, for which we have no cartographical evidence, the documents show that there were tenanted enclosures. In 1636 a tenant was instructed by the manor court to remove a hedge or fence and provide gates so that people could pass through the enclosure: 'A'paine laid that Margarett Stubb shall pull downe the fence in the parke land, and hang yates in the same, that men may passe as they have done heretofore ...' (Ingilby MS 1607, no. 5, 29 April 1636). An agricultural element and even tenanted holdings may always have been included, but at times parts of the park seem to have been totally converted to agricultural uses. In 1807, three years after Sir John Ingilby's return from exile, a commissioned map showed that the area designated as the park had contracted to comprise only the territory lying within about a quarter of a mile to the west and north-west of the castle. A park several times larger and was indicated by Chippendale in 1752, while the Tithe Map of 1838 shows a still larger recreational area, extending right to the medieval(?) park boundary earthworks on the western limits of the township. The reduced park existing at the start of the nineteenth century may have resulted from the crippling debts resulting from the dowry catastrophe, and it may also reflect its owner's enthusiasm for afforestation.

In 1813 an area larger than that mapped in 1807 yet considerably smaller than the medieval deer park at its greatest extent was defined and a wall that cuts across from old Whipley to Sadler Carr was built around it, enclosing the area that still comprises the deer park. George Elliot provided an estimate

> ... for getting stones, leading walling and finding lime and sand for the park wall beginning at low end where the other is finished to, and to go up the high-rails to near the house Occupied by Barney Wilks to be walled and finished in the same manner as the other it is joined to for the sum of ... per rood ... £1.2.0 total [£1.10]. The above to be finished

in a workmanlike manner at the satisfaction of a proper inspection by george Elliot Ripley (Ingilby MS 2550).

On the back is a note: '33/G estimate for Park Wall. This is done and he says it was a bad job'. Numerous patched sections at different points of collapse suggest that this was a shrewd observation.

Sir John's rebuilding work at the castle had produced a minor stately home that was widely admired. Its setting also attracted favourable comments, and during the course of the century efforts were made to develop the aesthetic potential of the park. A monument to the visual attractions of the place can be found in the form of the prominent earthworks of a prospect mount, conveniently located on high ground just to the north of the holloway of the medieval road to Whipley. Since vantage points such as this had a long currency, appearing in late medieval formal gardens, this mount is hard to date. It may well have been surrounded by a little garden of perfumed flowers and was sited to exploit a fine panorama of parkland. In the wood just to the west of the mount are the traces of an ice house, where ice from the lake below could have been stored for culinary uses.

The planting of the park is interesting and blends ornamental trees with ancient pollard specimens inherited from the landscape that existed before the park was created. Beech, fast growing but shallow rooted and prone to being blown over, were used, but the most appealing of the trees used in landscaping are the pollarded sweet chestnuts. These grow swiftly but are unable to produce ripened nuts in the English climate. Their timber has something of the appearance of oak, though not the durability, and from a distance they look very like the oak pollards that lined the lane to Whipley before emparking. A truly gigantic specimen that fell in gales in 1962/63 is still surprisingly well preserved and lies at the foot of the medieval quarries, where stunted ancient oak pollards are rooted amongst the quarry debris.

Land by the beck directly below the castle was probably under water from well back in the medieval period, as the Walk Mill place-name and several medieval references to mills show. The 1838 map shows a water body about 394 × 164 feet (120 × 50 metres) lying below the castle and directly to the south-west of the beck and it is captioned 'reservoir'. Shortly afterwards, ornamental lakes inundated the area of the old mill ponds and much more besides. The lakes were the creation of Sir William Amcotts Ingilby (1783–1854). On 5 June 1844, the contractor, M. Faviell, wrote to Sir William:

I beg leave to present you with the total expenses of your grand lake and have no doubt when you see the amount you will be satisfied. There are several parties to which I owe money yet such as the Blacksmith, Quarry men and Stone Masons but I think about £120 will pay all off excepting what you please to allow me for my time and Talent. I propose to be with you on Thursday night (if all is well) and in the meantime may if convenient look the accounts over and I will then endeavour to pay those men and so wind all up (Ingilby MS 2556).

FIGURE 59.
Historic features in Ripley Park, North Yorkshire

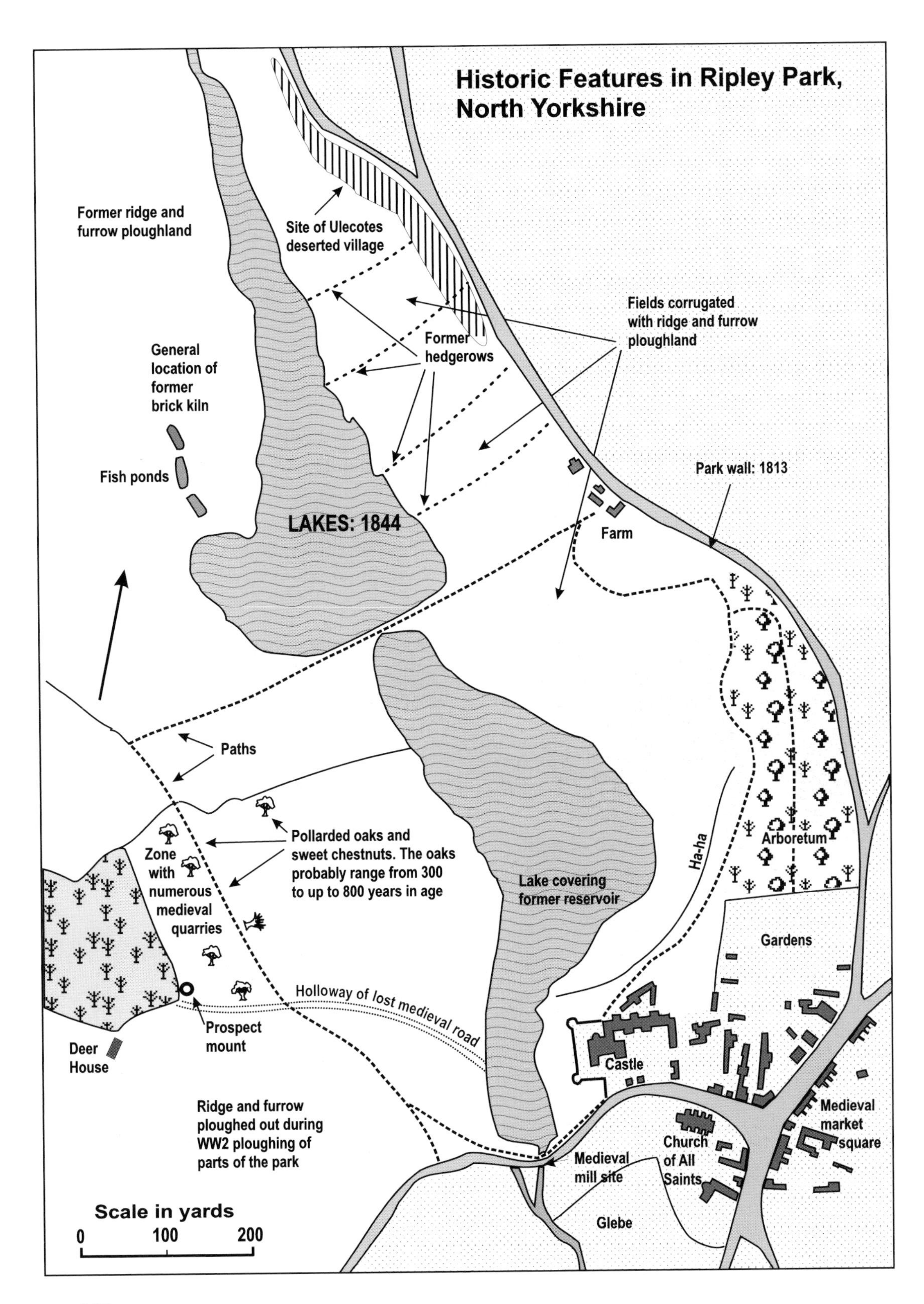
Historic Features in Ripley Park, North Yorkshire
Former ridge and furrow ploughland
Site of Ulecotes deserted village
Fields corrugated with ridge and furrow ploughland
Former hedgerows
General location of former brick kiln
Fish ponds
LAKES: 1844
Park wall: 1813
Farm
Paths
Pollarded oaks and sweet chestnuts. The oaks probably range from 300 to up to 800 years in age
Zone with numerous medieval quarries
Arboretum
Ha-ha
Lake covering former reservoir
Gardens
Holloway of lost medieval road
Prospect mount
Deer House
Castle
Ridge and furrow ploughed out during WW2 ploughing of parts of the park
Medieval mill site
Church of All Saints
Medieval market square
Glebe
Scale in yards
0
100
200

The lakes cost £3,000 and a footnote on the receipted bill, which was dated 13 June 1844, records: 'Settled this account, which was not to have exceeded £2,700, but I may be well clear of it at the above cost of £3,000' (Lady Alberta's red book).

Shortly after his work on the lakes, Faviell was employed to drain the park. In April 1848 he was paid £254 11*s.* 6*d.* (£254.57½) for laying 9,450 drainage tiles and spreading £10 worth of grass and clover seed (Ingilby MS 2561).

There is a local tradition that the landscape park, with its somewhat serpentine lakes, was the creation of Lancelot 'Capability' Brown. 'In the manner of' might be more accurate. The last of Brown's works that I can trace was Woodchester Park in Gloucestershire, begun in 1782. His designs involving lakes became fashionable in the 1770s, while in 1783, he died. Any plans by Brown for landscaping with lakes at Ripley must have gathered many layers of dust before the lakes materialised in 1844!

A late intrusion in the park was the Dower House built at Broxholme by Mary Anne, the widow of Sir Willam Amcotts Ingilby on land that he had bequeathed to her before his death in 1854. With ten bedrooms, stabling for eight horses and a lodge house, this was a mansion in its own right.

The refurbished castle and the landscape park provided the essential trappings of aristocracy and involvement with game and hunting had always been regarded as a hallmark of nobility. The winning of a right of free warren

FIGURE 60. A part of the old upland common at Ripley. The field boundary on the left was that of a very old intake.

and the construction of a deer park had marked stages in the aristocratic progress. Warrens were first constructed under the Norman kings to accommodate what were then known as coneys, with the rabbits imported from southern Europe being ill-adjusted to the cool, damp English climate. Both as culinary delicacies and as feeble creatures ill-equipped to survive in the wild, the medieval rabbits required careful conservation. The practice of keeping rabbits in warrens for their meat and pelts persisted long after the animals had become acclimatised and had proliferated. It is not known when the Ingilbys established a warren in the Scarah locality to the north of the park and the deserted settlement of Owlcotes but a Scarah Warren is marked there on the Jeffreys' atlas of Yorkshire of 1772. Nine years earlier there had been a dispute, recorded, but not very legibly, as follows: 'disputes have arisen between ? Hodgson ... to ? Iverson at Cayton, and William Kirby warrener to Sir John Ingilby at Scarah warren, touching the right of driving and chasing the rabbetts out of the grounds of the said ? Hodgson bounding upon the said warren ...' (Ingilby MS Additional Accession 2662). The warrener was found not to be at fault. Ploughing appears to have destroyed any structures, but from the description and from examples surviving elsewhere the warren may have consisted of one or more cigar-shaped 'pillow mounds' containing artificial runs for the rabbits to shelter and breed in, set within a densely hedged enclosure. Perhaps Hodgson had been jumping on a pillow mound in the hope of frightening the rabbits out?

Judging from the proximity to the Knox Mill, Scarah warren was probably the same as the Knox warren, mentioned in 1668–89 in connection with a case of trespass due to be heard at York assizes between Sir William Ingilby and one Ralph Leedam, who was accused of taking rabbits from Knox warren (Ingilby MS 3163). Poaching would remain a problem so long as there were hungry countryfolk, while the emergence of Harrogate as a fashionable spa and resort patronised by guests with a taste for game cannot have made matters easier for the keeper and his master. One fellow seems to have been a particular nuisance, leading to the posting of notices at the Ingilby estates:

> To Mr William Hebden, Take Notice, that I do hereby discharge you, from hunting, shooting, setting, coursing, fishing, or on any pretence coming upon my lands in occupation, in the township of Ripley, within the parish of Ripley, in the west-Riding of the county of York. If you shall presume to come thereon, after this notice, you will be prosecuted, as a wilful trespasser (Ingilby MS 2630)

Enclosure

Parliamentary Enclosure was a movement involving the fragmentation and privatisation of common land. It normally took place on a parish-by-parish (or township) basis and while in theoretical terms it was intended to increase the efficiency of farming by removing antiquated communal practices, in practice it favoured large landowners and disadvantaged the poor. Most parishes contained an element of the rural poor at the base of the social hierarchy, people who eked-out their existences by exploiting their access to the varied resources of the commons. When the commons went, the poorest people soon followed. In Nidderdale as a whole Parliamentary Enclosure exerted a very powerful effect on landscape and society, most importantly through the Enclosure in 1770 of the extensive commons of the Forest of Knaresborough, which extended across many parishes. Here, the old open, communal countrysides were partitioned and overlain by an angular network of straight walls and hedgerows with the new field and property boundaries being drawn on the maps of the surveyor and then superimposed upon the living landscape.

In Ripley the changes were less extreme. Firstly, the main ploughlands and meadows had been partitioned two or three centuries earlier by piecemeal processes of agreement between respective landowners, producing hedgerow patterns that snaked across the countryside to preserve the old selion boundaries. Secondly, since young John Ingilby, a junior, would inherit the greater part of the land in the township and since the allocations of land under Parliamentary Enclosure were proportional to each owner's holding in the un-Enclosed territory, when Scarah Moor (also known as Skalwray Moor) was enclosed in 1778, he emerged with 332 of the 384 acres (134 of the 155 hectares) of the old common.

On the eve of the Enclosure, the common existed as a somewhat diamond-shaped territory on the elevated ground in the north of the township. The main accessway came through the shrunken village of Birthwaite, and just beyond the place where trackways funnelled outwards into the common there were two large intakes. Their surrounding walls are of a construction that should predate the enclosure walling of the eighteenth century and the larger boulders at the base of the walls imply rebuildings on medieval footings. Before Parliament was petitioned for an Act of Enclosure, an agreement was reached between the representatives of the Ingilby family interests and the other landowners concerned, Walter Vavasour, Thomas Grimston and Elizabeth Eteson, with arrangements for the compensation of the lost rights in the common (Ingilby MS 2589). A written description of the proposed partitioning of territory was provided and it provides glimpses of the old countryside of assarts surrounding the common

> ... the said Walter Vavasour in Conson of his having assigned to him a piece or parcel, of Ground part of the said Common containing by Admeasuremt: forty seven acres bounded by his ancient Inclosure in the Township of South Stainley and the ancient Inclosures late of the said John Ingilby in the Township of Ripley called Ruddings and Darnbrough pasture commencing at the South Corner of a Close belonging to the said Walter Vavasour in Stainley called Vivars Close then in the Occupation of Joseph Matson and ending at the South West side of the Fence

> [a wall or hedge] of Darnbrough Pasture in an Exact Line from the North East Corner of a Close called Barron Garth and on the South West by the same Close to a certain old Pollard Oak standing in the Fence thereof and so ranging in an oblique Line westward ... (ibid.)

Agreements were reached concerning the fencing or the new holdings and making of roads and Vavasour, Grimston and Eteson were given the right to divert water from Slate Ridge Well stream to water their cattle.

The costs of expressing the agreement in the landscape were considerable; the Act of Parliament cost £195 and the total cost of Enclosure came to more than £1,387, though as Jennings notes, it was more than compensated by a rise of more than £100 a year in the Ingilby rental (1983, p. 333). The expenses were as follows (Ingilby MS 2589–90)

1,480	roods of fence wall, 5ft. 9ins. High	£446 3 0
858	roods of ditching and quicks*	57 3 4
1,440	posts and rails	72 0 0
340	double posts and rails	26 0 0
40	posts and 120 rails	4 12 0
29	gates, 58 oak posts, ironwork	21 15 0
100	roods of carriage road making	30 0 0
		£657 15 4
3	sets of barns and other buildings	£535 0 0
		£1,192 15 4

* quicks are quickthorns – living plants, commonly hawthorn, used for hedging.

The village

Ripley was not the only planned medieval village to undergo a complete rebuilding in the nineteenth century and East Witton in Wensleydale, rebuilt by the Earl of Aylesbury in 1809, was another. However, Ripley is surely the first planned medieval village to undergo a relatively modern rebuilding by the same family that created it. Sir William Amcotts Ingilby (1783–1854) was the man responsible, a wealthy politician with a humourous and eccentric manner: 'I have been a wicked rascal in a certain way, like many others, but I hope I've never injured or oppressed any one' (Sir Thomas Ingilby, p. 17).

In recreating Ripley on a grand scale, around 1827, Sir William was helped by the wealth from the Lincolnshire estates of the Amcotts family that he had inherited in 1807 and by the sale of 300 acres of famland (the retention of which would have been more sensible as it was situated in what would become the core of the expanding spa of Harrogate). At first he began to rebuild in a conventional, vernacular-based style but then departed for a European tour. It is said that it was then that he was inspired by an example he saw in Alsace Lorraine to build a model estate village of his own. This example from the

Franco-German marchlands is said to have been the inspiration for the deep continental overhanging eaves that feature in the new village architecture. However, it is impossible to imagine that Sir William could have lived much of his life in England without being thoroughly aware of a well-established tradition for enhancing the approaches to a mansion or park with a custom-built estate village of a memorable design. Every time he travelled from Ripley to Leeds he would have gone through Harewood, built for Edwin Lascelles by Carr of York. The West Riding also contained the Moravian settlement of Fulneck; in the North Riding there were Brandsby, Hackness and East Witton, and in the East Riding there were New Sledmere, Langton and Settrington – all of them existing before Ripley was rebuilt, while Amcotts country in Lincolnshire had Harlaxton, Normandy and Well Vale. At the time when Ripley was being built, the vogue for 'villages of vision' had reached its peak and work was in hand at a good number of other English sites. If the architecture of Ripley is influenced by continental motifs, one could equally well argue that it is essentially a home-grown Tudor revival style.

By 1830, the entire village had been rebuilt. Tenants exchanged dwellings that must mainly have been centuries old for substantial stone-built homes complete with bedrooms heated by cast iron fireplaces bearing the crests of the Ingilby and Amcotts families. The community was served by a general store and the shops of a butcher and a tailor, a smithy, three inns, the Boar's Head, the Oak and the Star, and public laundry facilities. The three inns did not survive for their centenary. In the second decade of the twentieth century another Sir William Ingilby became concerned that his tenants were too enthusiastically engaged in Sunday drinking and so he imposed a ban on Sunday opening – with the result that the three landlords decided to quit the village. There was no pub again until the opening of the Boar's Head in 1990. (After being thrown by his horse, which was startled by a child dashing from one of the village houses, Sir William banned the villagers from using their front doors (Sir Thomas Ingilby p. 20). The Free School, founded by Mary Ingilby in 1702 was demolished and rebuilt along with the rest of the village. Fortunately for the landscape historian, accurate maps portray the situation before and after the rebuilding. The Chippendale survey of 1752 and the map of 1807 by Calvert and Bradley of Richmond (Ingilby MS 2517) show the situation before rebuilding, with constricted streets, the gaps in the roadside frontage and the little pinfold where trespassing cattle were held at the northern margin of the village. The tithe map surveyed by James Powell of Knaresborough in 1838 shows the more ordered lay-out of the model village. As described in 1866:

> ... it has frequently been pronounced by travellers one of the prettiest little places in the North of England. The town was rebuilt in 1827–8, in the Tudor style of architecture, by the late Sir William Amcotts Ingilby, Bart., and consequently has a very modern appearance. It consists of one principal street, extending North and South, with a minor one

leading to the Castle. The houses are all commodious and handsomely built – even the cottages have an appearance of more than ordinary comfort and respectability. The streets are spacious, and bear an aspect of healthiness (Thorpe, 1866, p. 2).

Although the pattern of property divisions was completely obliterated and redesigned in the course of rebuilding, the fundamental lay-out of the settlement survived remarkably unchanged. Thus the legacy of the medieval episode of village creation was preserved and carried forward in the next episode of village planning, more than four centuries later. From my point of view the most intriguing aspect of the entire affair concerns the question of whether any of the buildings erected by the medieval village founder remained standing until his distant descendent pulled them down about 1827. From the recollections of the character of these dwellings as recorded by Thorpe in 1866, this seems quite likely.

The settlement structure provided by the street pattern was unchanged, though the streets were widened during the rebuilding: '... the streets being little more than half their present width, and instead of being macadamised, they were laid with round pavestones' (Thorpe, p. 97). In the surrounding countryside the old houses surviving from the sixteenth and seventeenth

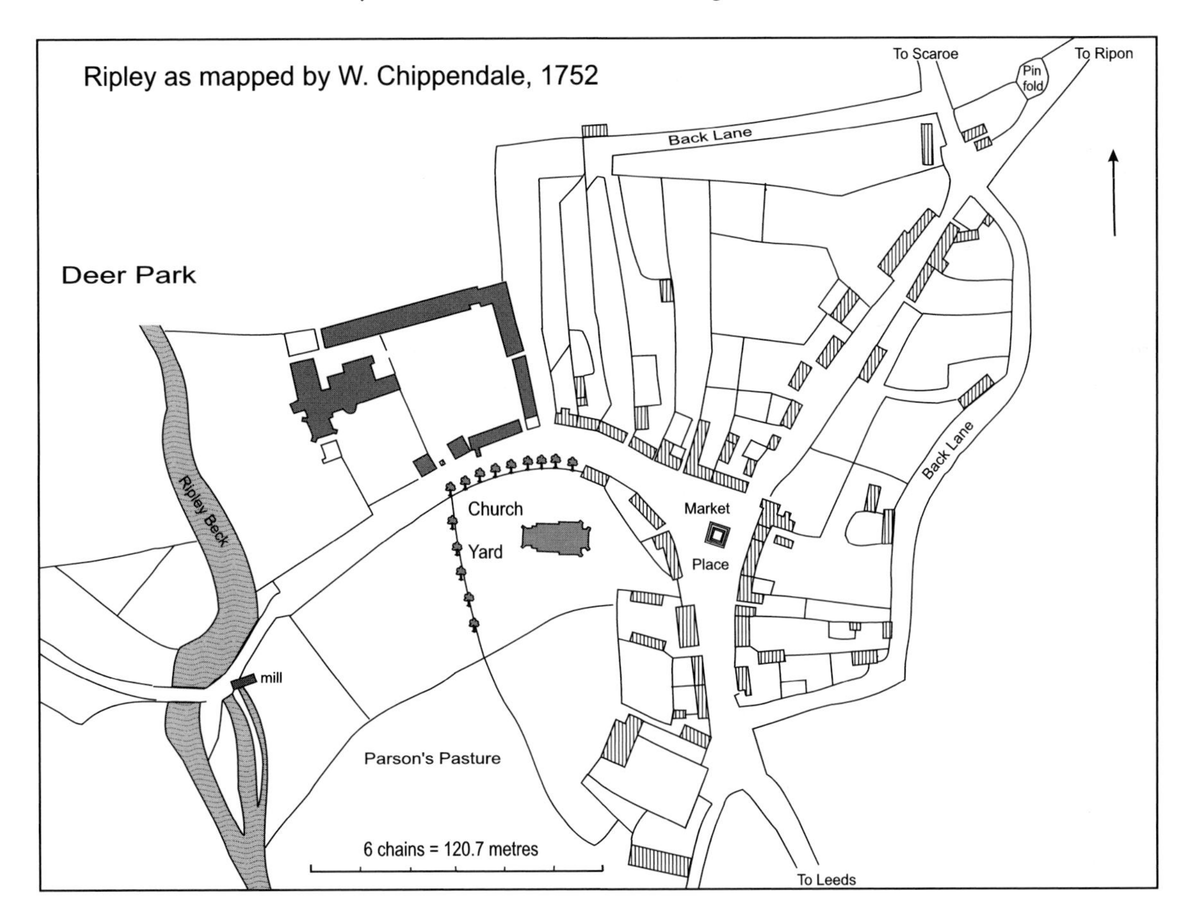

FIGURE 61. Ripley, 1752: as mapped by W. Chippendale. Note the mill on the beck. The streets are narrower and the buildings less organised than after the rebuilding.

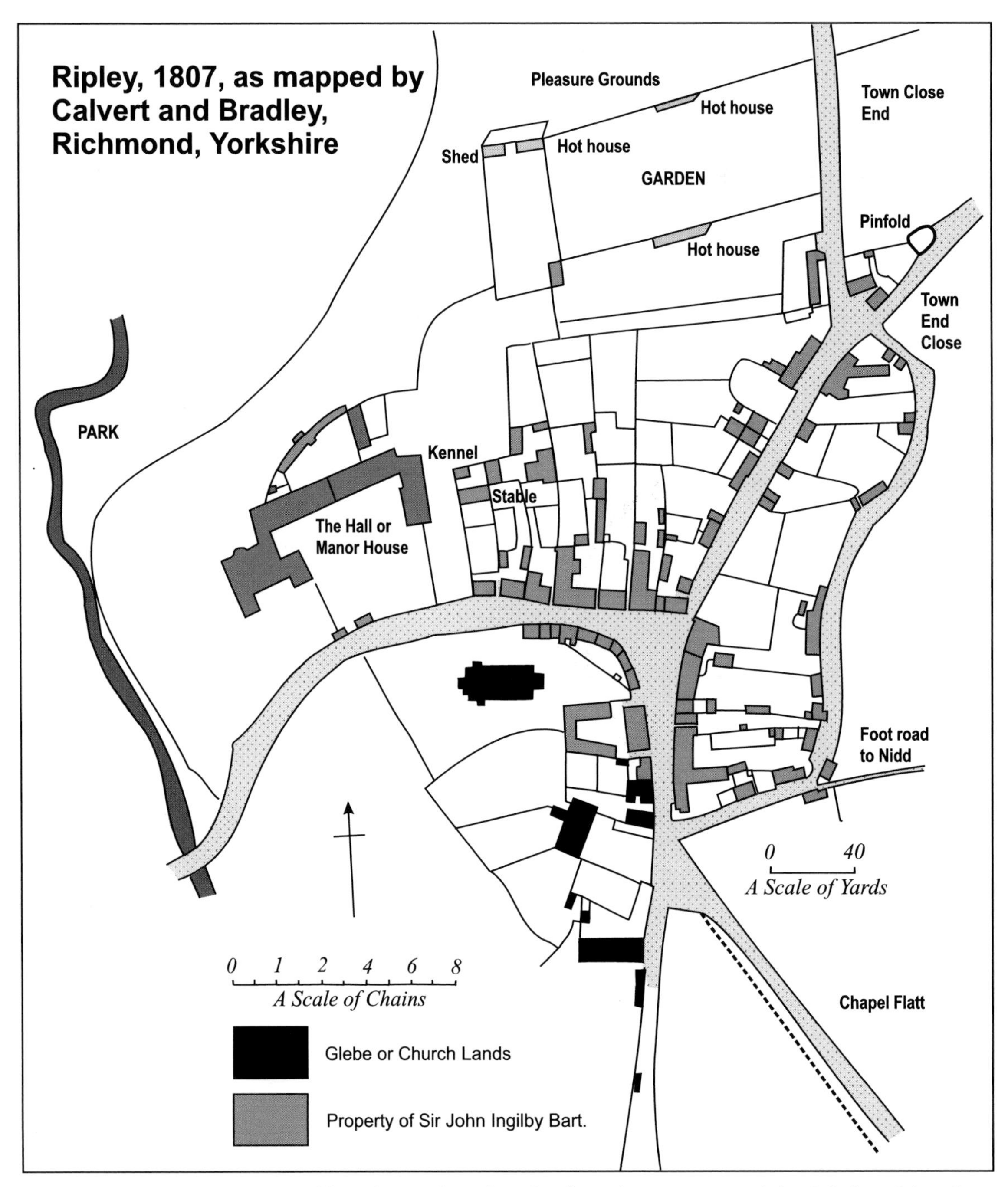

FIGURE 62. Ripley, 1807: as mapped by Calvert and Bradley of Richmond. An accurate and detailed plan of the village a little before its rebuilding. Note the pinfold in the north-east, where livestock that had strayed were impounded.

FIGURE 63. Ripley, 1838: as surveyed for the tithe map by James Powell of Knaresborough. This map shows the village shortly after rebuilding – essentially the settlement that is seen today. Note the field name 'Walk Mill Ing', which recalls a medieval fulling mill.

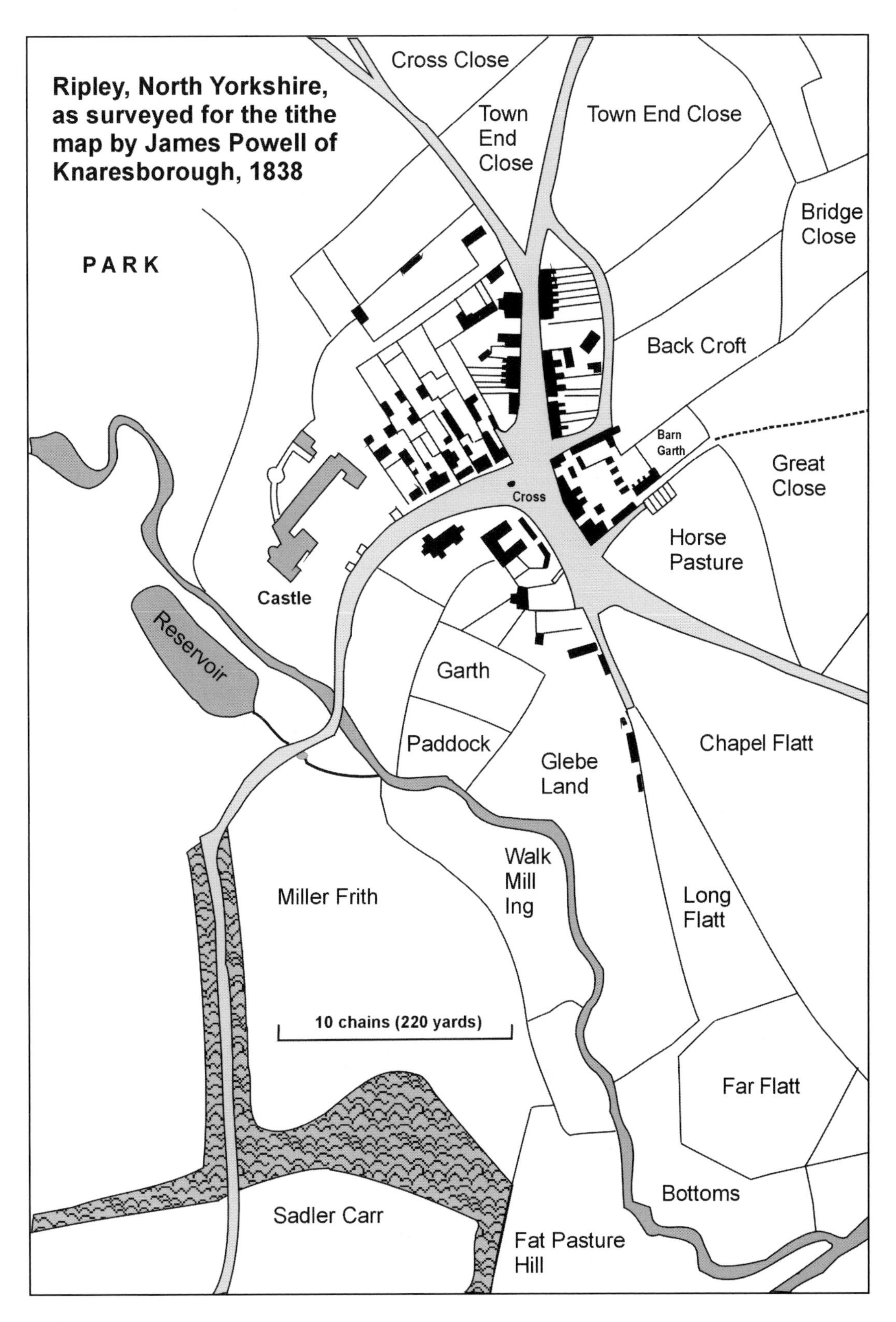
Ripley, North Yorkshire, as surveyed for the tithe map by James Powell of Knaresborough, 1838
Cross Close
Town End Close
Town End Close
PARK
Bridge Close
Back Croft
Barn Garth
Great Close
Cross
Horse Pasture
Castle
Reservoir
Garth
Paddock
Glebe Land
Chapel Flatt
Walk Mill Ing
Miller Frith
Long Flatt
10 chains (220 yards)
Far Flatt
Bottoms
Sadler Carr
Fat Pasture Hill

centuries tended to be of a cruck-frame construction, with pairs of blades fashioned from split, curving oak boughs being used to form the 'A'-shaped frames that defined bays and gave the main support for the thatched roof. In old Ripley, however, most of the houses were taller, two-storey structures, many apparently having an upper storey of timber, lath and plaster carried upon a lower storey of stone. Such a building, thatched and with its front door apparently opening to a lobby backed by the base of its chimney which had fireplaces for the rooms to either side, is shown standing facing the church across the market place in one of the pictures commissioned by Sir John Ingilby in the 1780s. It is hard to draw conclusions from such a detail in a painting, but the design of *this* house, at least, would seem to be no earlier than the seventeenth century and the building could well have been erected in the middle of that century.

Thorpe tells us that in old Ripley

> The houses chiefly had a very ancient appearance, which at once indicated a remote origin; they mostly consisted of two storeys in height, the lowermost or basement being entered by a descent of two, three, and sometimes four steps, and the low roofs were almost invariably covered with thatch, with the upper room windows in the gables; several of these dwellings being of the half-wall and half-stucco kind, that is, the outer walls of the lower storey being masonry and the upper ones of lath and plaster (p. 97).

He described cottages on the Back Lane, which contained various ruined old buildings as being similarly entered down several steps and he also mentioned

FIGURE 64. Houses deriving from the nineteenth-century rebuilding lining the cobbled area of the former market at Ripley.

FIGURE 65.
Examples of the nineteenth-century architecture influenced by Tudor styles.

FIGURE 66.
Pollards, daffodils and the 'village of vision' architecture at Ripley.

the traces of a road with dwellings along its left side that had led towards the site of the original church. Facing the buildings was a row of fine elms, the decayed trunks of which were still visible in 1866. A huge elm had stood at the southern entrance to Ripley (we are not told on which side of the track) while a still larger tree, its hollow trunk, 16 feet in circumference, had stood near the centre of the village until blown down in a gale in January, 1839. He mentions that some old habitations had been converted into cattle sheds, but without specifying their locations. Then there was the market cross; stocks can still be seen at the stepped cross base, but this is not the original stocks: '... formerly the old Parish Stocks used to stand here at the South front of this Cross, so that the criminal undergoing punishment could sit on the first step'.

One building, a bizarre addition to the orderly yet still rather rustic landscape, was not constructed until 1854. Whatever the provenance of the domestic architecture, the *Hotel de Ville* a sort of village 'town hall' was a eccentrically continental addition. The style is Gothic revival and within the energetically soaring lines of the exterior was a worthy interior contained a lecture hall, baronial in style, post office and accommodation for the staff, reading-room, library and Mechanics' Institute. Apart from the mid-twentieth-century by-pass, this was the last significant addition to the village landscape. The last link with the ancient countryside must have been shattered with the arrival of the eleven and a half mile branch line linking Harrogate and Pateley Bridge (one should say 'near arrival', for following the level margins of the Nidd floodplain, it missed the various villages by considerable margins). Some of the leading figures in Nidderdale, like the Greenwoods at Birstwith, enthusiastically favoured the line. The Rev. Sir Henry John Ingilby (1790–1870) opposed the railway as an intrusion upon his domain, but faced with the prospect of compulsory purchase he agreed to cut the first sod when the track was laid. The first service pulled into Ripley station on May Day, 1862.

The philosophers of geographical study have talked about 'time/space compression', with distances shortening as the pace of change intensifies. In the span of my lifetime, I may experience more change than took place in the entire medieval period. Similarly, the Victorian villagers who saw the railway track being laid through Ripley might well have been effectively closer to their medieval forbears lying beneath the plough soil above the abandoned church than they are to ourselves. They all knew when to sow barley, how to deal with docks or charlock and the causes of foot rot in sheep. Few of today's villagers will have a clue about such things.

The age-old bonds between villagers and countryside is withering away. The lands that once allowed life to go on are now only scenery. The surnames in the local telephone correspond less and less with those on the old tombstones in the churchyard. Families depart or die-out and the gaps in the settlement pattern are filled by outsiders buying in to a rural myth of timeless traditions, homespun wisdom and wholesome fare. As the quality of history, now glossily packaged and marketed for Middle England, declines, so the appetite for it increases. The members of the rootless classes cast about anxiously for roots and identity in a world of vacuous clamour. Time and place, history and site, provide moorings for them. The countryside no longer exists to feed the countryfolk; supplies can be bought more cheaply from elsewhere. Instead, it seems to exist to serve as a context for 4x4 lifestyles that provide sufficient status and role-playing opportunities for those whose fortunes were made elsewhere. But countryside has gone on through everything and will always go on. Each society has left its distinctive impress upon the scene. How much of the evidence of the older, more substantial rural lifestyles will survive when this countryside yields to the next, one cannot know.

Landscape is not just countryside or scenery, it is recorded history and social commentary. Each society creates landscape in its own image and several successive communities were involved in crafting and assembling what we see at Ripley today. Throughout the world we are still creating landscape in the image of our society – which strongly suggests that there will be less of strength and character to captivate future generations of landscape archaeologists. This is not the first account of Ripley, and certainly not the last word on the subject. But as time passes it will be harder and harder to employ the techniques used here. I have often been conscious that it would have been far easier to be working at the time of Jeffreys, around 1770, or Grange, in the 1860s, when far more evidence will still have been surviving intact. Enough remains to make the little township a very special place indeed.

What I have written is almost all that I know about Ripley Township. Over

FIGURE 67.
Ripley village: the church and castle are towards the bottom right.
© ANTHONY CRAWSHAW

a century from now, another researcher may begin a new investigation, drawing on my efforts as I drew on the recollections, insights and follies of the Victorians. I hope the charms and fascination of the place may be but little diminished.

References

Sir Thomas Ingilby, Bt, *Ripley Castle.*

Sir John Ingilby, letter to the Somerset Herald [he is an officer not a newspaper], 22 December 1784, quoted in Sir Thomas Ingilby, Bt *Ripley Castle.*

Jennings, B. (ed.), *A History of Nidderdale*, Huddersfield, 1983.

Pennant, T., *A Tour from Alston-Moor to Harrowgate and Brimham Crags*, John Scott, London, 1804.

Sheeran, G., *Landscape Gardens in West Yorkshire*, Wakefield Historical Publications, Wakefield, 1990.

Thompson, M. W., *The Decline of the Castle*, Cambridge University Press, Cambridge, 1987.

Township of Clint-cum-Hamlets, Women's Institute of Burnt Yates, Ebor Press, York, 1982.

Thorpe, J., *Ripley: Its History and Antiquities*, Whittaker, London, 1866.

Weater, W., *Some Historic Mansions of Yorkshire*, Richard Jackson, 1888.

Index